T0282651

CAMBRIDGE LIBRARY COLLECTION

Books of enduring scholarly value

Earth Sciences

In the nineteenth century, geology emerged as a distinct academic discipline. It pointed the way towards the theory of evolution, as scientists including Gideon Mantell, Adam Sedgwick, Charles Lyell and Roderick Murchison began to use the evidence of minerals, rock formations and fossils to demonstrate that the earth was older by millions of years than the conventional, Bible-based wisdom had supposed. They argued convincingly that the climate, flora and fauna of the distant past could be deduced from geological evidence. Volcanic activity, the formation of mountains, and the action of glaciers and rivers, tides and ocean currents also became better understood. This series includes landmark publications by pioneers of the modern earth sciences, who advanced the scientific understanding of our planet and the processes by which it is constantly re-shaped.

An Account of the Basalts of Saxony

Jean François d'Aubuisson de Voisins (1769–1841), geologist and engineer, was an Officer of the Légion d'Honneur, Knight of St Louis and Chief Engineer at the Royal Mining Corps. He published numerous papers on geology, mining and hydraulics, and is best known for his textbooks, *Traité de Géognosie* (also reissued in this series) and *Traité d'Hydraulique*. He studied geology and mineralogy in Freiburg with Abraham Werner, the key proponent of Neptunism, the theory that all rocks had an aqueous origin. Later in his career Daubuisson was to side with the Plutonists, who argued that basalts formed from molten rock. However, in this paper, published in French in 1803, he describes his observations of the basalts of Saxony and argues that they, and all basalts, are sedimentary. This English translation by the Secretary of the Wernerian Natural History Society was published in 1814, and provides a fascinating insight into this discredited but once influential theory of the Earth.

Cambridge University Press has long been a pioneer in the reissuing of out-of-print titles from its own backlist, producing digital reprints of books that are still sought after by scholars and students but could not be reprinted economically using traditional technology. The Cambridge Library Collection extends this activity to a wider range of books which are still of importance to researchers and professionals, either for the source material they contain, or as landmarks in the history of their academic discipline.

Drawing from the world-renowned collections in the Cambridge University Library and other partner libraries, and guided by the advice of experts in each subject area, Cambridge University Press is using state-of-the-art scanning machines in its own Printing House to capture the content of each book selected for inclusion. The files are processed to give a consistently clear, crisp image, and the books finished to the high quality standard for which the Press is recognised around the world. The latest print-on-demand technology ensures that the books will remain available indefinitely, and that orders for single or multiple copies can quickly be supplied.

The Cambridge Library Collection brings back to life books of enduring scholarly value (including out-of-copyright works originally issued by other publishers) across a wide range of disciplines in the humanities and social sciences and in science and technology.

An Account of
the Basalts of Saxony

With Observations on the Origin
of Basalt in General

J.F. Daubuisson
Edited and translated
Patrick Neill

CAMBRIDGE
UNIVERSITY PRESS

CAMBRIDGE
UNIVERSITY PRESS

University Printing House, Cambridge, CB2 8BS, United Kingdom

Published in the United States of America by Cambridge University Press, New York

Cambridge University Press is part of the University of Cambridge.
It furthers the University's mission by disseminating knowledge in the pursuit of
education, learning and research at the highest international levels of excellence.

www.cambridge.org
Information on this title: www.cambridge.org/9781108048422

© in this compilation Cambridge University Press 2013

This edition first published 1814
This digitally printed version 2013

ISBN 978-1-108-04842-2 Paperback

AN

ACCOUNT

OF THE

BASALTS OF SAXONY,

WITH

OBSERVATIONS ON THE ORIGIN OF
BASALT IN GENERAL.

BY

J. F. DAUBUISSON,

MEMBER OF THE NATIONAL INSTITUTE, AND ONE
OF THE PRINCIPAL ENGINEERS TO THE
BOARD OF MINES IN FRANCE.

TRANSLATED, WITH NOTES, BY

P. NEILL, F. R. S. E. & F. L. S.

SECRETARY TO THE WERNERIAN NATURAL
HISTORY SOCIETY.

WITH A MAP OF THE SAXON ERZGEBURGE, FROM PETRI.

EDINBURGH,

PRINTED FOR A. CONSTABLE & CO.,
AND LONGMAN, HURST, REES, ORME & BROWN,
LONDON.

1814.

PREFACE

BY THE TRANSLATOR.

THE question whether the formation of
basalt is to be ascribed to the agency of
water or of heat, has long been agitated
among the mineralogists of the Continent;
those who maintain its igneous origin being commonly denominated Vulcanists or
Volcanists, while the title of Neptunists
has been bestowed on their opponents.
The former consider the conical summits
of basaltic hills as the craters of extinct
volcanoes, and the tabular or columnar
masses of basalt, as the remains of lavas
which have flowed from them. The Neptunists, on the other hand, contend, that

a

basalt has had a common origin with all other rocks, and that the whole have been deposited by water. It is the doctrine of the Continental Volcanists which Mr DAU-BUISSON chiefly combats in the following Memoir.

The Volcanic theory has been defended with great ability by VOIGT, BREISLAK, CORDIER, FAUJAS ST FOND, and others on the Continent; the late Sir WILLIAM HAMILTON taking the lead among the Volcanists of England.

To the Neptunian standard were attached BERGMAN, KARSTEN, and KIRWAN, — names deservedly remembered with veneration: and Dr RICHARDSON has also acquired reputation as a Neptunist, by his papers on the celebrated basaltic district of the north of Ireland *. But the doctrines of the illustrious WERNER hold unquestionably the first place among the opinions connected with the

---

* *Transactions of Royal Irish Academy;* and *Phil. Trans.* for 1808.

aqueous theory of basaltic and other rocks.
These doctrines, early diffused throughout
the Continent of Europe by the just celebrity of their author, were first introduced into Britain by Professor JAME-
SON ; and from him they have been adopted by two of the most distinguished sys-
tematic writers on Chemistry in this
country, Dr THOMSON and Mr MURRAY.

A sort of intermediate opinion concerning basalt seems to have been received by
some eminent mineralogists, namely, that
a substance possessing all the usual characters of that rock may be produced, according to circumstances, either from fusion or from aqueous solution. Such was
the view of this question entertained by
DE SAUSSURE, and ultimately by DOLO-
MIEU. Such likewise were the sentiments
of SPALLANZANI ; and VON BUCH, as
well as BRONGNIART, incline to the same
opinion. Even the author of the following
essay, Mr DAUBUISSON, after visiting
Auvergne, became satisfied that some of
the rocks *of that country*, which, in external characters, closely resemble common

basalt, have been produced by the action of volcanic fire. It has indeed been alleged, that he has changed his opinions to a much greater extent, and now considers all basalt as of igneous origin. This, however, seems to be a mistake : it is possible that he may not now agree with Werner in considering volcanoes as confined to the Newest Flœtz-trap formation; but his conviction concerning the aqueous origin of the basalts of Saxony appears to remain unshaken *.

In Scotland, a more general question, involving the problem of the igneous or aqueous origin of basalt, has of late years been keenly debated between the supporters of the theory of the late ingenious Dr Hutton of Edinburgh, and the disciples of the Freyberg Professor. The Huttonian theory, however, differs, in certain particulars, very materially from the Volcanic; and these variations are of such a nature, as to furnish the advocates of Dr Hut-

---

* *Posteà*, Note u. p. 266.

ton's opinions with answers which are
more than plausible, to arguments that are
quite conclusive against the Volcanist. By
the former theory, for example, it is not
necessary, in order to account for the presence of basalt, to search for the ruins of
a crater from whence that substance may
have been ejected, or to trace the remains
of an ancient current of lava ; for it is not
supposed by that theory, that every basaltic hill is an extinct volcano : basalt is
considered as the product of a fusion which
took place " deep in the bowels of the earth,"
or " several miles under the surface of the
ocean *," where not only the access of atmospheric air was prevented, but a degree
of *compression* existed, sufficient to hinder
the escape of elastic fluids; so that the melted substance might remain quite compact
or free of pores, and calcareous matter
might be exposed to great heat without being converted into quicklime,—or bitumi-

a 3

---

* *Illustrations of the Huttonian Theory,* 8vo. pp. 21,
30, 69.

nous matter without being changed to coke.
Basalt is thus supposed to be a kind of
" subterraneous or unerupted lava," produced by the action of heat under a great
degree of pressure. The followers of Dr
Hutton, therefore, as remarked by Mr
Playfair, are rather to be viewed as Plutonists than Volcanists.

It being a preliminary, indispensable to
any argument concerning the origin of Basalt, that the meaning of the term should
be determined with precision, the author
of the following memoir has devoted to
that object, one of his introductory sections. For this determination, in the present state of mineralogical science, it would
appear that it is only requisite to attend
to the character of the *mineral species*
entitled Basalt in the system of Werner;
the arrangement of that eminent naturalist, so far at least as *rocks* are con-
cerned, having justly superseded every
other. There is reason, however, to imagine, that too little attention has often
been paid to the method by which Werner,
in the establishment of his species, has

supplied the want of that individuali-
ty of character which so greatly facilitates
the arrangement of animals or plants.
In Werner's method, the principle of spe-
cification is founded, not on any selection
of the appearances or properties of minerals, but on the *suite of external characters* which they present, ALL of which are
to be attended to in the discrimination
of the species. It is evident, therefore,
that, for their correct determination, a *suite
of specimens* is indispensable ; of specimens
shewing the different appearances of the
rock, or taken from different parts of the bed
or vein which it constitutes. It may very
easily happen, that, in small detached specimens, it may be difficult to point out
the difference between rocks which, on the
great scale, are perfectly distinct ; between
grey-wacke, for instance, and some varieties of sandstone, mica-slate and gneiss,
or between sienite and granite : But the
examination of a suite of specimens can
scarcely fail to remove every difficulty.
It may further be observed, that although
Werner has depended principally on the
external characters, he does not, for the

a 4

determination of his species, confine himself to them alone ; but likewise employs,
particularly what he terms the *geognostic
character*, that is, the relation in respect to
position, which the mineral bears to the accompanying substances : And Professor
Jameson remarks, that the geognostic character is very important, and regrets that
it has hitherto been so little attended to *.

While, therefore, we have the evidence,
not only of the Volcanists above named,
but of Von Buch and Daubuisson of the
Wernerian school, that a substance completely resembling basalt may be produced
by a volcano ; it is not to be inferred de-
cisively, from any thing at least advanced
by the last-mentioned authors, that if a
suite of specimens be examined, and especially if reference be also had to the *geognostic* character, the true nature of the
substance may not be determined.   It
may thus be very true, that if particular

* *System of Mineralogy*, edit. 1804, vol. i. Introd.
XXXV.

specimens, taken from the Vesuvian lava
of 1794, (remarkable for its similarity
to basalt,) or from the extinct volcanoes
of Auvergne, be compared with particular specimens from the basaltic summits of Germany or Scotland, no difference
shall appear, sufficient to enable us to dis-
criminate between them, far less to point
out their distinctive *specific characters.*
But it is evident that the rules laid down
by Werner, will not, in this instance, have
been complied with: and if we have re-
course to a series of characters, illustrated
by a suite of specimens, it can scarcely
happen, that a practised mineralogist will
find difficulty in distinguishing the true
basalt of Werner from the *basaltic lava* of
Vesuvius, Auvergne, or any other volcanic country. If, further, the geognostic
character of the substance be ascertained, all difficulty must in general cease;
and, in a strictly correct view, we cannot even speak of the Basalt of Werner without having reference to its natural alliances.

The chain of reasoning by which the
solution of the question respecting the ori-

gin of basalt, is connected with the general theory of the earth, appears to be the
following.—If basalt be any where observed gradually to pass into other minerals, the aqueous origin of which, is not
disputed ; if it any where exhibit evident
marks of stratification, or alternate with,
and pass into rocks of a distinctly stratified structure, or which contain petrifactions ; no doubt can remain concerning
its origin in that particular instance : and
if it be shewn that, universally, the basalts
of every region of the globe agree in characters, and in geognostic relations, it seems
impossible to refuse to assign to all of them
a common origin.

It is admitted by the distinguished author
of the Illustrations of the Huttonian Theory*, that the origin of Basalt, Greenstone,
Amygdaloid and Wacke must be the same,
the change from the one substance to the
other being made through a series of insensible gradations.   In the following memoir, the gradual passage of the rocks com-

---

* *Illustrations*, &c., 8vo. p. 67.

posing the trap suite in general into each
other, is established ; of clay into wacke ;
wacke into basalt and amygdaloid ; or of
slate-clay into Lydian-stone, and of this into basalt; of basalt into greenstone or green-
stone-porphyry ; and of greenstone into
clinkstone-porphyry- In this view, the
question is not merely about basalt, but includes the origin of all the rocks now enumerated ; and these, grouped together in the
manner here stated, are diffused as widely
over the crust of the globe as the most distinctly stratified rocks, sandstone and limestone, which are universally admitted to
be of aqueous formation : And it may be
added, that limestone passes gradually into wacke, in the trap-tuff of Fifeshire ;
and that quartzy-sandstone passes in the
same way into greenstone, at Salisbury
Craig near Edinburgh.

With these distinctly stratified rocks,
basalt and greenstone, in beds of great extent, and uniform thickness throughout,
are sometimes observed to alternate : a
fact explained by the Volcanists, by supposing, that volcanic eruptions and sub-

marine deposites have taken place al-
ternately; and which the Huttonians account for, by alleging that the basalt or
greenstone have been forcibly interposed
between the beds of sandstone or limestone
subsequently to *their* formation. But it
not unfrequently happens, that basalt and
greenstone, considered separately, exhibit
marks of stratification; and when this is
the case, even Faujas St Fond admits that
they must have had an aqueous origin.

Professor Jameson, in his Lectures, has
recently stated some new views on stratification, as connected with trap-rocks. He
observes, that in some of these, small portions of limestone, slate-clay, and quartzy-
sandstone occur, so intermixed, as to prove
the simultaneous formation of the whole:
in other cases, portions of limestone and
slate-clay assume the form of small alternating layers, in the trap; and it not unfrequently happens, that beds of limestone,
slate-clay, and clay-ironstone, several hundred yards in extent, may be observed alternating with each other, but still included in one vast bed of trap; the whole as-

semblage, consequently, in Mr Jameson's opinion, is to be considered as of contemporaneous origin, or the result of a single deposition *.

Mr Daubuisson proves from different striking facts and appearances, that basalt must have been deposited *from above :* he remarks, that the crystals of olivine, augite, and other substances considered as extraneous, are in general equally and uniformly dispersed through the mass of the basalt,—a thing not likely to have happened in a confused and violent eruption : and he shews that basalts brought from countries the most remote, are perfectly similar in character, and contain the same kind of ingredients ; while the lavas even of contiguous volcanoes differ very materially in character and in constituent parts.

---

* In this view, the sixty-five alternating beds enumerated in Note F. pp. 217—220. as occurring on the coast of Fife, may possibly be reduced to a few, or even to one great contemporaneous deposition.

Having given a very general view of the important questions illustrated and discussed in the following memoir ; I shall subjoin a brief account of the author *.

Mr C. F. DAUBUISSON is a native of the south of France, and nearly related to the celebrated General ANDREOSSI, formerly Ambassador to the Court of Britain. He received the rudiments of his education in France, and was early distinguished for the extent and accuracy of his mathematical knowledge, and the acuteness of his views in other branches of science. He afterwards resided upwards of six years in the Electorate of Saxony, and diligently studied mineralogy and the art of mining at Freyberg, under Werner. To the latter study in particular, in all its branches, both of theory and practice, he devoted a very

---

* For whatever information I possess concerning the Author, I am indebted to the friendly communications of Professor JAMESON, who spent several years at Freyberg, and was personally acquainted withhim.

considerable portion of his time, and with such remarkable success, that when, in 1802, he left Freyberg, a place which may be regarded as the metropolis of miners, he was considered as thoroughly informed in that art. Being fully convinced of the importance of a knowledge of mineralogy in the practice of mining, he devoted himself also to that science, with his characteristic zeal and ardour; and, after some years of constant application, in the lecture-room, the closet, and the field, proved himself well entitled to the appellation of a skilful mineralogist.

The first mineralogical essays of Mr Daubuisson, were published in the *Journal de Physique* during his residence at Freyberg. These were principally descriptions of simple minerals, and of rocks, drawn up from the lectures of Werner, and were probably intended merely to make his countrymen better acquainted with the system of that mineralogist. At this time, also, he contributed liberally to the *System of Mineralogy* of BROCHANT, then printing at Paris: And no translation having appear-

ed in France of WERNER's *Treatise on
Veins,* first printed in 1791, he published
at Freyberg, in the year 1801, under the
immediate inspection of the author, an excellent translation of that valuable work.

In the year 1802, while he was still in
Germany, but on the eve of leaving Saxony, he published at Leipsic, in 3 vols.
8vo., an *Account of the Mines of Freyberg ;*—unquestionably one of the most
agreeably written and popular books on
the art of Mining, that has hitherto appeared.

Soon after his return to France, he
made several journeys into the different
departments of that empire, with the view
of making himself acquainted with thè
practice of mining in his native country.
During these expeditions, he acquired much
information concerning the French practice, and communicated to the Directors of
Mines his views of the superior advantages of the German system of mining, as
conducted by Werner, to whom Europe
is highly indebted for the improvement

of this art. On his return, he drew up descriptions of several of the French mines, which are inserted in the *Journal des Mines.* His descriptions of those of Huelgoat and Poullaoüen are excellent, and may serve as models for similar reports.

He had been appointed to one of the inferior situations in the mining department; but his merits having been thus speedily disclosed, he soon rose to the high rank of " Ingenieur en chef au Corps Imperial des Mines."

He appears always to have been warmly attached to the Wernerian *oryctognostic* system, having written several papers expressly in explanation and support of it, in different periodical works*.

He has also stood forward as the champion of his Master, in answer to an attack

b

---

* Particularly in the *Journal de Physique,* t. 54. & 60.; and *Annales de Chimie,* t. 57. & 62.

of Mr CHENEVIX. That distinguished
chemist, after spending about a year at
Freyberg, in the study of mineralogy, appears to have taken a disgust at the Ger-
mans, and, in different French and English Journals, amused himself with decrying
them, both as a nation and as philosophers.
A principal attack was levelled against
Werner and his mineralogical system, and
the passages on which it was founded,
were chiefly attempted to be derived from
the writings of Daubuisson. This called
forth from our author a spirited answer *,
in which he proved that Mr Chenevix had
both misunderstood and misquoted Werner, and had thus the merit of erecting
himself, the fabric which he had undertaken to destroy.

Fully aware of the importance to geology of an accurate measurement of
heights, particularly by means of the ba-
rometer, Mr Daubuisson examined the
various methods at present employed in

---

* *Annales de Chimie*, t. 69.

such mensurations, and has given his views on this subject, in an excellent paper in the *Journal de Physique* for 1810.

His *geognostical* memoirs, although few in number, are valuable. They are all contained either in the *Journal de Physique,* or the *Journal des Mines.* Perhaps the best is that entitled, " Statistique mineralogique du department de la Loire," in the latter Journal for April 1811.

Mr Daubuisson has lately published in France several accounts of simple minerals. One of the most important is " on the hydrate of iron considered as a mineral species." Another, scarcely less curious, is on Clay-slate ; in which we have the first analysis published of that important mineral.

The " Memoir on the Basalts of Saxony," of which a translation is now laid before the public, was read before the National Institute in July 1803, and published soon afterwards as a detached volume. In 1804, the author read before the Insti-

tute, the paper already alluded to, on the
country of Auvergne : an abstract of which
has appeared in the foreign journals ; but
the paper itself has not yet been published, or at least it has not yet reached this
country.

In executing the translation now presented to the public, I have endeavoured
to convey the Author's meaning accurately, and at the same time with freedom ; but
I am very sensible that many blemishes
will require the reader's indulgence.

I have added some short explanatory
notes in the course of the Memoir, and a
very few Notes of Illustration at the end ;
all of them being distinguished by the let-
ter T. (Translator.)

Where the height of mountains, or the
distance of places are spoken of, I have
reduced the French *myriametres* to English
miles, and the *metres* to feet. It nowhere
happens that any argument is founded on

the exact measurements specified, else I should also have given the French expression, to avoid ambiguity or inaccuracy.

The original publication was not illustrated by any map : and on the Continent, where the *Erzgeburge* or range of Metalliferous Mountains, is well known to all who interest themselves in mineralogical pursuits, a map was perhaps unnecessary : but in this country, such an aid, I thought, was indispensable ; no fewer than fifteen important mountains of Saxony, Silesia, or Bohemia, whose summits are capped with basalt, being described in the Memoir, and a good deal of the argument turning on their relative position with respect to each other, and to the mountain-ranges with which they are connected. Through the obliging attention of Dr Fitton (now of Northampton), and of Mr Arrowsmith of London, I therefore procured a correct extract from the large map of Saxony, in fifteen sheets, constructed from surveys made by the Prussian engineer Petri, in the years

1759, 1760, 1761, and 1762. This extract contains the greater part of Upper
Saxony, and a part of Bohemia and Silesia, including of course the entire mountain-chain of the Erzgeburge, and also that
of the Mittelgeburge ; and considerable attention has been paid, to mark particularly those basaltic mountains or eminences which are described in the first part
of the Memoir, and the origin of which
constitutes the chief topic of inquiry ; so
that the reader will be enabled to find the
position of almost every place mentioned
in the work. I hope, therefore, that the
map will not only render the descriptions
and reasoning of the book more perspi-
cuous and instructive, but will not be without more general interest, from the mineralogical celebrity of the district which it
represents.

I have in general adopted the language
of the Wernerian School, as employed
in the writings of Professor Jameson. In
a branch of knowledge, but lately introduced in this country, it must on all hands
be admitted, that new terms were abso-

lutely required ; and if, in those which have been adopted, elegance or congruity with the genius of the English language have in a few instances been sacrificed to expressive distinctness, there can be no objection to a future reform. But they who have most loudly complained, have not hitherto offered any substitute whatever for the terms which they condemn ; and some of the most obnoxious of these, begin to creep even into the writings of the strongest opponents of Werner.

P. N.

CON-

ANALYTICAL

TABLE OF CONTENTS.

Part II.—Of the Basalts of Saxony.

Part III.—Inferences respecting the formation of the Basalts of Saxony.

Part V.—Inferences respecting basalt in general.

Account of the Properties of Basalt.

NOTES.

A

AN

ACCOUNT

OF THE

BASALTS OF SAXONY, &c.

INTRODUCTION.

A MARKED diversity of opinions on any subject connected with physical science, cannot long continue; the very existence of different sentiments, tending to lead to a knowledge of the truth. When, in the course of discussion, the facts come to be verified, and
precise and definite meanings are affixed to
the terms employed, any dispute about the
conclusions must soon be settled. The con-
troversies which, at various times, have taken
place, about the Newtonian discovery of gravita-
tion,—the figure of the earth,—the measure
of forces, or, more recently, on the subjects

of combustion and oxidation, may be mentioned as examples. The question now agitated
among mineralogists concerning the origin of
Basalt, will probably have a similar result, at
no very distant period.

§ 1. The following is the state of the question on this subject

Some naturalists, led by curiosity to visit
the volcanoes at present in a state of activity, had remarked, that the rock-masses
composing the mountains in which these volcanoes are situated, and which are themselves
the products of eruptions, are of a blackish
colour; that they often resemble scoriæ;
that they contain particular foreign ingredients; and occasionally shew a tendency to
the prismatic structure. Upon travelling into
other countries, not reputed volcanic, they
found mineral substances having a considerable similarity to those volcanic products;
possessed of the same sort of dark fuliginous colour; containing the same kind of foreign materials; occasionally with a vesicular structure,
recalling the idea of scoriæ; and frequently

columnar: they found these substances composing hills and mountains of a conical shape,
like Vesuvius and Etna. Such traces of
resemblance having strongly impressed the
imaginations of these travellers, they boldly
proclaimed the discovery, in those other countries, of ancient volcanoes, where in former
ages subterranean fire had produced all its
terrible ravages, but where it has long been
happily extinguished. The abundant black
rock-masses in such countries, (chiefly basalt),
were of course considered as volcanic productions, and pronounced to be lavas.

Other naturalists, after exploring the same
mountains accurately, and with impartial
coolness, perceived that their entire mass
was not composed of basalt, or the supposed
lava; but that, while this substance formed a
covering over them, their main body was actually of the same nature as the rock of the
surrounding country; that no appearances indicated any excavation or rending of the interior; but that their mass was solid and continuous throughout, and very far indeed from
being a confused heap of fragments of different

stones, scoriæ, or melted materials, as is the
case in true volcanic mountains. After examining with the greatest care the nature of
this basalt, they were of opinion, that it possesses such strong affinities to other rocks,
(particularly wacke and greenstone*,) the origin
of which is not doubtful, that it should be regarded merely as a modification of such rocks.
They remarked, that the basalts of all countries,

---

* The *wacke* of Werner is a substance intermediate
between basalt and clay. It contains less iron, and
perhaps more lime and magnesia, than basalt; its
texture is less compact, and it decomposes more
readily. Its other properties may be seen in Brochant's
Mineralogy, vol. i. p. 434. The popular meaning of
the name *wacke*, is different; it is bestowed by the
Saxon miners on all mineral substances, which do not
contain metallic ores, especially if such substances be
of a dark colour.

Grünstein, or Greenstone, is a rock with a granular
structure, composed of grains of hornblende and felspar,
larger or smaller in size, and of a lamellar texture.
The hornblende generally prevails in the mixture; it
is of a blackish-green colour, and often tinges the felspar, which is commonly whitish. Greenstone is one
of the rocks which Werner comprehends under the
title of Trap-rocks.—Brochant, vol. ii. p. 581, & 606.

have a close resemblance to each other; that basalt
is every where the same mineral, containing
the same sort of foreign substances, possessing
the same external form, and general aspect, in-
dependently of the rock on which it is incumbent. Among the foreign ingredients which
it contains, some could not undergo the action
of fire, without being destroyed or entirely
changed. In some places it has been found to
include petrifactions,—the remains of organic
beings: in others, it rests on beds of coal,
which are not in any degree altered or affected
at the points of contact. In other places still,
it alternates, in beds, with limestone, sandstone, and even coal; and appears therefore to
have had a similar origin. The general thin-
ness of the beds of basalt, their great extent,
and their bendings, do not accord with the notion of their having arisen from a stream of melted matter. The prismatic, and sometimes vesicular or porous structure of basalt, and the conical shape of basaltic mountains, may be observed in many other kinds of rocks. Such
arguments, together with a number of others,
to be afterwards stated, have led those mine-

ralogists to entertain doubts concerning the
volcanic nature of basalt, and to think that no
reason appears for supposing its origin to have
been different from that of the other rocks
which, along with it, constitute the solid crust
of the globe.

Geologists are thus divided into two parties,
on the subject of the formation of basalt; the
one ascribing it to fire, the other to water.
They bestow on each other reciprocally, the
whimsical titles of *Volcanists* or *Vulcanists,* and
Neptunists, from the name of the heathen deity
under whose banner they seem to be ranged.

Some observers, (De Saussure and Dolomieu
among others), who having perhaps too easily
yielded to first appearances, had embraced the
former of these two opinions, have latterly
been staggered by the arguments of the Neptunists. After having examined attentively the
strong facts brought forward, they have in some
measure departed from their former opinions,
and have admitted, that there exist certain
kinds of basalts which never have undergone
the action of fire. Here then is an approach to
the Neptunian views: and I am persuaded, that
if correct observers, possessed of the necessary

physical and mineralogical knowledge, were to employ themselves on this interesting question, the difference of opinion concerning the formation of basalt, would soon be decided.

2. It is evident, therefore, that this question can be resolved only by observations made with equal care and discernment. I myself have been led by particular circumstances to examine a great number of basaltic mountains. Besides, in returning lately from the mines of Silesia, I traversed, on foot, and with my mineralogical hammer in my hand, almost all the basaltic hills of that range, which extends from the Carpathian Mountains, as far as the banks of the Rhine. But my particular attention was directed to those of Saxony. A long residence amidst these mountains, had made me familiarly acquainted with them. I had examined them at different times; and on the spot discussed the results of my observations with other mineralogists: I was thus enabled to supply deficiencies in my first observations, and to review whatever needed elucidation. I flatter myself, therefore, that I am able to produce numerous and well

attested facts, which may perhaps furnish
some data for the solution of the problem.
The purpose of this essay, is to make these
facts known: and if I shall be led to conclude, that the basalt of Saxony is not of
volcanic origin, I by no means pretend finally
to decide the question as to basalt in general.
I claim only the right enjoyed by every one
who publishes his observations, that of giving
his own opinion on the subject of which he
treats, and assigning his reasons for adopting it.
I expressly disclaim any right to pronounce positively on things which I have not seen with
my own eyes, although I may sometimes be
induced to form conjectures from analogy.

3. In order to obviate any doubt as to the
meaning of terms, I shall begin with explaining what is here to be understood by *basalt*,
and by the expression *volcanic production*. Then,
after giving a concise sketch of the position and
nature of the chain of mountains, I shall particularly describe each basaltic eminence in succession. I shall next mention the conclusions
which seem to me to result from my observations, particularly in regard to the formation of

the basalts of Saxony: stating my reasons for believing that they are not, and cannot be, of volcanic origin. After having taken a cursory view of the extent of the great basaltic mass which constitutes a part of the surface of the globe, I shall hazard an opinion concerning the nature of basalt in general; and then conclude with an examination of the volcanic hypothesis of that excellent geologist DOLOMIEU, whose misfortunes lately excited so much interest among the learned in every part of Europe, and whose recent loss is still generally deplored *. These different matters will form the objects of the five parts into which this essay is divided.

---

* It will be recollected, that Dolomieu, who spent much of his time in investigating the volcanoes of Italy and Sicily, was unexpectedly and cruelly cast into prison, by order of the Court of Naples. He remained immured in a dark and loathsome dungeon in Messina, for nearly a year, notwithstanding the intercession of Sir Joseph Banks, and other distinguished characters; till Buonaparté, having attained to the First Consulship, effectually interposed in his favour, and obtained his release. He did not long survive his enlargement, sinking under an attack of fever in winter 1802, in the 51st year of his age.—These are the circumstances alluded to by the author. T.

———————

PART I.

PRELIMINARY DEFINITIONS.

I AM now to explain with precision, what I mean by the term *basalt*, and by the expression *volcanic production*.

4. If the mode of formation of basalt were to be taken into consideration, it might be possible to give an accurate and satisfactory definition : but as this would be taking for granted a thing which is precisely the point in dispute, I must wait until I have established its origin.

Not being yet prepared, therefore, to give a *definition* of basalt, I shall give a description of it, or detail its characters and properties. Few minerals possess characters so distinct; and few indeed are so easily recognized : all basalts, even of countries the most distant from each other, have a common resemblance, and are not like to any other mineral; and they may all be easily comprized under one description.

BASALT.

The colour of basalt is greyish-black, more or less deep. When it is polished and moistened on the surface, it has a bluish aspect. In some varieties, the colour has a tint of green; in others of brown or red: In the former case, the basalt approaches to wacke or to greenstone; in the latter, to common clay-ironstone.

It occurs massive; in beds covering mountains; of which it generally forms the summit. Frequently these masses or beds are divided into prisms or columns more or less regular in structure, and which are generally several yards in length: sometimes they are divided into tables, or into balls with concentric layers. Some varieties present numerous vesicular cavities.

Its fracture is dull and almost earthy, but fine-grained; sometimes it passes into the large and perfect conchoidal, at other times in-

to the coarse-grained uneven. It often presents
granular distinct concretions *. It is difficultly frangible, when not traversed by rents.
The prisms, especially when they are small, resound under the hammer, as if it had struck the
anvil. The fragments are more sharp-edged, the
nearer that the fracture approaches to the conchoidal; and the conchoidal fracture obtains
in the hardest and most compact kinds.

Its hardness varies. The varieties in which
the fracture is conchoidal, give some sparks
with the steel: the others may be scratched
with the knife: Those which abound with
vesicles, often feel dry and rough to the touch.

Its specific gravity is about three times that
of water; 3,065 according to Klaproth. But
as the weight diminishes in proportion to the

---

* See Brochant's Mineralogy, vol. i. p. 118. for an explanation of this expression, and of some others bor-
rowed from the language of the Wernerian school.   D.

The reader may be referred also to Professor Jameson's Treatise on the External Character of Minerals;
his technical phraseology being generally adopted in this
translation.   T.

number of vesicles which it contains, some
fragments are much lighter.

Almost all basalts affect the magnetic needle:
But Haüy has shewn that only particular points
in the basalt are endowed with this power.

Exposed to the action of the atmosphere and
elements, it decomposes more or less readily into a rich blackish earth, well suited to vegetation. The very compact and iron-like varieties,
however, seem to resist all decomposition.
Those which are vesicular, often decompose into a kind of greyish powder, which sometimes
resembles ashes.

Subjected to the action of heat, it melts, according to Sir James Hall, at about 38⁰ of Wedgwood's pyrometer: it then changes into a
brownish-black or greenish black glass, a little
translucent on the edges. If this glass be
again melted, and cooled very slowly, it once
more assumes a stony aspect. Exposed to
heat in a crucible, covered with powdered charcoal, the basalt changes into a grey dull mass,
full of little cavities containing globules of
iron.

Both M. Klaproth and Dr Kennedy analysed
basalt; and found, that in a hundred parts it
contained,

	Klaproth.	Kennedy.
Silica, ...	44.50	46
Alumina, ...	16.75	16
Oxide of iron, ...	20.00	16
Lime, ...	9.50	9
Magnesia, ... , . . .	2.25	0
Soda, ...	2.60	4
Water, ...	2.00	5
Oxide of manganese, ...	0.12	0
Muriatic acid, ...	0.05	1
Loss, ...	2.23	3
	100.00	100

Klaproth observed likewise a trace of carbon;
and in the 5 parts of water in the analysis
by Dr Kennedy, are included the gaseous substances.

Such are the properties which characterize
and distinguish basalt taken by itself. It may
here be observed, that in giving the name of
basalt to the mineral which possesses these
characters or properties, I have in view the
same stone to which the ancients gave that
name. Pliny, speaking of the different species
of marble, mentions one which the Egyptians

brought from Ethiopia, and which they called *basaltes*, on account of its hardness and colour resembling those of iron. " Invenit eandem " Ægyptus in Æthiopia, quem vocant basal- " tem, ferrei coloris et duritiæ; undè et no- " men ei dedit." Lib. 36. cap. 7.

In order to give a complete account of basalt, I ought to consider it in its geognostic relations;—to give some details concerning its structure and form, mention the mineral substances which generally accompany it, and state the peculiarities of its *repository*. But a knowledge of these circumstances not being necessary for characterizing basalt taken by itself*, and belonging rather to *its* geological history, I have placed them in a note. (Note A.)

5. Under the denomination of *volcanic productions*, I include only those substances which have been completely melted, and have had their nature changed by a subterranean fire, and which have been subsequently placed in the

* The soundness of this opinion may be doubted. See Translator's Preface. T.

situations where we now see them, by a volcanic eruption: In short, I mean substances similar to those lavas which Vesuvius and Etna continue to pour out at this day. I thus exclude those fragments of minerals, which the force of a volcanic eruption may have merely torn from one place, and by ejection deposited in another; and likewise those which have been only altered in some degree by the action of heat, such as a shivered crystal.

As I myself have never had an opportunity of visiting active volcanoes, I shall not pretend to enter on any explanation of their origin, of the phenomena which they present, or the effects which they produce: at the same time, I may be excused for exposing the hypothetical and arbitrary suppositions which have been formed on the subject of volcanoes, and for treating with perfect freedom those causes and agents of a sort altogether peculiar*,

---

* Alluding in particular to the *principle*, different from caloric,—the " vehicule quelconque," by which Dolomieu supposed the matter in the interior of the earth to be maintained in a state of fluidity analogous to fusion, but materially different from igneous fusion. See Note X.   T.

which have been so gratuitously admitted by
Dolomieu; and the pretended existence of
which, far from being supported even by probability, is directly contradicted by all that we
know of physical certainties. As he took for
granted the existence of facts which were not
to be found in Nature, it was to be expected
that he should attempt their explanation by
imagining causes not subject to her laws.

As specimens of lava for comparison with
other rocks, I cannot admit any but the produce of Vesuvius, Etna, or of other active and
well known volcanoes. I must reject all arguments founded on specimens taken from extinct volcanoes, whether the existence of those
volcanoes have been real or imaginary; because
their existence is the very point in dispute.

I certainly regard the fluidity of lavas
as an effect of caloric; and I consider their
heat as sufficiently intense to melt stony
matters, and to keep them in a state of fusion;
for I have no reason to believe the contrary:
According to all that I have learned from persons who have seen the flowing of lavas, the
liquid torrents spread around them a bright

light; and they set fire to, and consume, every combustible material which they meet with in their progress. These are just the effects of ordinary fire. The homogeneous nature of the substance of lavas, proves, that there has existed sufficient heat completely to melt the mineral bodies which enter into their composition.

But it may be alleged, that people often walk on the crust which is formed on the surface of the lava, while the stream still continues to flow beneath; and that, through the fissures of the crust, is seen the light which proceeds from the melted matter; yet that but little heat is felt. There is, however, nothing extraordinary in this. I have frequently observed the same thing in our foundries. I shall mention only one instance: At Freyberg, they smelt ores, which in a hundred parts of earthy matters, do not contain one of metal. They mix the ores with iron pyrites, and scoriæ from former lead-works. The lower cavity of the furnace, contains from seven to eight quintals (hundred pounds) of melted matter, and it is about two feet in

diameter. When the furnace is broached, the
matter in fusion which flows into this cavity,
produces so much heat, that no one can approach; but in a very short time, it becomes
covered with a crust, through the crevices of
which may be perceived the flame of the sul-
phur of the pyrites, which continue ignited
below. This crust is not more than three or
four lines in thickness: it can bear, however, a
considerable weight; and it intercepts the heat
so much, that I have at such a time been able
to make experiments, without being incom-
moded by it. In a quarter of an hour, the
crevices become blocked up, and one may pass
close by the bason, without perceiving that it
is full of metallic and stony matters in a state
of fusion; but as soon as the crust is lifted up,
and the fluid substances exposed, the heat
becomes so powerful that a person must re-
cede many steps. As to what is alleged about
copper medals having been enveloped in the
lava, without being melted, I must have
witnessed all the particular circumstances
attending the fact, before I could be able to
draw any conclusion from it. It would, among

other things, be necessary to know if the lava had been in immediate contact with the metal, and what was its precise nature.

I have not, it is true, ever witnessed volcanic eruptions; but I have seen a number of subterranean fires; and it is probable that those of volcanoes are of the same nature. The subterranean fires which I allude to, are nothing else than certain thin beds of coal, which are burning some yards under ground. There must necessarily, no doubt, be a great difference between the effects of this trifling cause, and those which are produced by volcanic agents. These last may be expected to be proportioned to the vastness of the cause. However, the subterranean fires first mentioned, throughout the extent of their sphere of activity, vitrify clays and earthy substances; melt the surface of masses of quartz ; and roast ironstone, causing the reduced metal to flow out. I have been an ocular witness of all these facts*, and have seen them not farther than ten paces from the place where the beds of coal

---

* Note B.

were still burning. All the burnt-clays and porcelain-jaspers which abound in the cabinets of mineralogists, are only clays semivitrified, or which have undergone complete vitrification, in these pseudo-volcanoes. If, then, the heat produced by an inflamed bed of coal, which extends only to the depth of some yards under the surface of the earth, is sufficient to melt quartz and ironstone; can we believe, that the heat of volcanoes is not of such intensity, as to fuse copper, if fairly exposed to it? or is it possible to conceive, that a stream of melted stones, of more than three hundred feet in thickness, should not at all affect or alter a bed of bituminous matters, over which it should flow * !

PART II.

OF THE BASALTIC MOUNTAINS OF SAXONY.

§ 6. Before describing the individual basaltic mountains of Saxony, I shall shortly mention

* See the description of Mount Meisner, in Note C.

the position, the extent, and the general nature of the mountain-chain to which they belong.

The chain is called the *Erzgebürge*, or Metalliferous Mountains. It separates Bohemia from the electorate of Saxony. Its direction is from N. E. to S. W. One of its extremities terminates in Franconia, where its base is joined to that of the Fichtelgebürge. On the other side, it ends in the great and deep valley occupied by the river Elbe; this valley separates it from another chain which lies between Bohemia and Lusatia, and which, having nearly the same direction, may be regarded as a prolongation of it. This last chain joins the mountains of Silesia.

The length of the *metalliferous chain*, is about 120 miles, in the line of its direction. Its height is about 3280 feet, above the plains of Saxony, and 3600 above the level of the sea. The declivity towards Bohemia, is very rapid; but towards Saxony it is quite gradual. The whole chain presents the appearance of an amphitheatre of mountains, the tops of which, are either rounded or flattened. This chain gradually rises from the level at the foot of the amphitheatre, through a tract of nearly thirty-seven miles, to the summit of the ridge. The declivity which looks towards Saxony, exhibits, through its whole extent, cultivated fields, meadows, and forests.

The fundamental rock of the chain, is granite. The granite is covered, and as it were wrapped round with beds of gneiss, of mica-slate, and of clay-slate; placed over one another in the order in which they have here been named. In many places, the granite seems to pierce the covering, and appear at day. Among these beds, some are found which contain metallic minerals. These, as well as the numerous and rich veins which traverse them, are the objects of the great mining operations of Saxony. There occur likewise in the chain, rocks of serpentine, and of quartz; beds of limestone, of coal, of clay, and others. The whole of the eastern part of the chain, is covered, on the north side, with a huge bed of porphyry, and on the south

side, with a bed of sandstone, of equal magnitude *.

7. On a chain of hills of this structure, does
the basalt rest, of which I am to treat. Under
various shapes, as tables or platforms, cones, and
domes, it forms the summits of about twenty
mountains, some of which are isolated, but
which more generally are connected by their
sides to the neighbouring mountains, the basaltic top alone remaining separate. This is
commonly the most elevated point in the neighbourhood; so that when, in surveying the aspect of any part of the chain, an isolated summit is seen rising above the surrounding mountains, we may almost conclude that it is composed of basalt. It is principally in the neighbourhood of the ridge of the chain, that mountains with basaltic summits occur. Precisely
on the highest part of the chain, between
Gottesabe and Irgand, there is a kind of mountain-plain, more than three miles in length,

---

* For more minute details on this subject, see
Daubuisson on the Mines of Freyberg, vol. ii. chap. 1

the rock of which is basalt. The back of the chain, which looks towards Bohemia, presents a greater number of basaltic mountains; but I mean here to confine myself to those of Saxony. These are known under the names of Scheibenberg, Bærenstein, Pæhlberg, Heidelberg, Ascher-hubel, Landberg, Steinkopf, Lichtewalde, Geissengenberg, Luchauerberg, Cottanerspitze; and beyond the Elbe, are situated the Winterberg, the Heulenberg, and Stolpen. Perhaps, besides these, there may be some others which are not known to me. Fragments of basalt which are found in particular places, favour this supposition, though I do not know of any others having been observed. All the basaltic rocks of Saxony taken together, present a surface of little more than an English square mile, and they are dispersed over an extent of country of more than six hundred square miles, so that they do not constitute the six hundredth part of the superficies of the chain of which they form a part.

I shall now give a particular description of each of the mountains above named. I shall, in each case, begin by noticing the position

and aspect of the mountain; I shall next
mention the nature and direction of the beds
of which it is composed; and then examine
the circumstances attending the superposition
of the basaltic summit, specifying its form and
extent.   Afterwards I shall take a general view
of the structure of the basaltic summit on the
great scale; and shall conclude with some notice of the nature of the basalt of which it may
happen to be composed, and of the foreign
minerals which it may contain.   To avoid repeti-
tion as much as possible, I shall not treat of
any thing that is not peculiar to the individual
mountain, referring to § 4. for what is common to the basalt on their summits.   I shall
further suppress some local details which
might have been necessary, if my design had
been to give a complete description of these
mountains, and not to confine myself to the
instructive appearances which they present.
I must add, that, notwithstanding the pains
which I have taken, I have found it impossible
to examine with the eye, every thing that was
necessary to render each description complete.
In a number of places, the rock was covered

with soil, so that I could not observe the
mode of superposition, the structure, and other
circumstances: in these cases, I behoved to
content myself with examining the points
which were exposed; and I believe there
are few instances in which I did not exercise
particular care.

8. The mountain called *Scheibenberg*, is situated at the distance nearly of fifty miles to
the S. S. W. of Freyberg, and nine miles from
the summit of the chain. Its eastern base is
washed by a small river called Tschopau, on
which lies the town of Schletau: the little
town of Scheibenberg is built on the western
base of the mountain.

Its general form is that of an unshapely
truncated cone. Its summit commands all the
country for six miles around, being elevated
between 1000 and 1300 feet above the base.
The general acclivity of the sides is very gradual; and towards the top, the body of the
mountain presents a rounded back, forming a
kind of mountain-plain, on the middle of which
lies an enormous platform of basalt. The sides

of the mountain are covered with cultivated lands; the eastern side, indeed, with wood; and the summit is naked, producing only mosses and lichens.

The body of the mountain is composed of gneiss: this is covered by beds of mica-slate and clay-slate : On the side next Schletau, the gneiss is visible at the highest part of the body of the mountain; the mica-slate about the middle; and the clay-slate towards the base. The general inclination of the beds, especially near that place, is from 50° to 60° to the N. E. The mountain is traversed by metalliferous veins, which are worked in some places. Situated on the kind of plain formed by the highest part of the body of the mountain, we find a bed of gravel; over which, there is one of fine sand; and then one of clay. These substances lie in horizontal beds, the one above the other. The upper bed, in the place where I made my observations, was from sixteen to twenty inches in thickness; it consists of an ochry-yellow clay, very smooth and unctuous to the touch, and which is used in the manufacture of earthen-ware.

On these horizontal beds of gravel, of sand,

and of clay, lies the mass of basalt which crowns the mountain. It is about 750 feet in length, 400 in breadth; and its thickness may be estimated between 200 to 260 feet. Towards the north and east, its sides are perpendicular; towards the west, they have a considerable declivity; and to the S. W., there is a basaltic hillock or eminence. The upper surface is almost horizontal; it inclines only a little to the west: it is covered with fragments of basalt, and with mosses, and grass.

In a great part of the lateral surface of the platform, the basalt is quite exposed; it is divided into irregular vertical prisms, producing the appearance of a row of pillars, surrounding the summit. These columns are very long; some of them even 130 feet; their thickness is from three to six feet. The number of sides in the prism varies; but most commonly each has six. These sides, far from being smooth or straight, are full of bendings and sinuosities. The angles are blunted. Thus, these prisms have nothing of regularity in their appearance, but form unshapely columns, which are divided by rents or horizontal fissures, at certain distances.

In respect to oryctognostic characters, the basalt of Scheibenberg presents nothing particular: it is somewhat less black, less hard, and less compact, than that which constitutes the fine regular columns to be seen at Stolpen and other places.

It contains a great quantity of small crystals of amphibole or common hornblende. I have not been able to determine the form of the crystals, as they have been restrained by pressure during their formation: perhaps they should rather be called grains with a lamellar structure, than crystals. As the hornblende resists decomposition better than the basalt, the grains of the former, which happen to exist at the surface of blocks of the latter substance, present projecting points: this circumstance, joined to their deep-black colour in the midst of the greyish-black, of the decomposed basalt, gives to these blocks the appearance of a mass in which an infinite number of small fragments of charcoal is disseminated. I have likewise observed in certain portions of the basalt of Scheibenberg, some grains of common olivin.

On the west side, a little above the town, many subterraneous galleries have been opened, which have been pushed under the basaltic platform, in order to procure sand and clay. These operations have disclosed a part of the beds on which the basalt rests. In 1787, M. Werner observed in this place, a bed of wacke, on which the basalt was deposited in immediate contact; and he remarked, that these two substances passed into each other by gradual shades of difference. In publishing this discovery, he announced his opinion concerning the origin of basalt, and thus threw down the gauntlet. An animated debate ensued between him and M. Voigt *. I had no opportunity of examining the particular spot where Werner made his observations, the falling of some rocks having entirely covered the place with debris. The point of fact, however, has never been denied †.

---

* This controversy, I understand, was originally carried on in the *Bergmännishes Journal* of Freyberg, *Allgemeine Literatur Zeitung*, and Voigt's *Mineralogische und bergmännishe Abhandlungen*. Several of the papers were afterwards translated into the *Journal de Physique*.   T

† NOTE D.

On the eastern side of the mountain, there are numerous heaps of rubbish, evidently dug from the interior, as well as many vestiges of ancient galleries, which had been driven under the basalt. These facts evince that there had been, in former times, a great number of subterraneous works in that part; but there remains no evidence to shew what minerals may have been the object of these labours. In the middle of the basaltic platform itself, there is a gallery called the Dwarf's Hole. This gallery may be entered to the distance of about 130 feet; but the water which occurs, hinders farther access. I have here observed small pieces of basalt entirely softened; while others which were contiguous, resisted the hammer, and were very hard. Two pieces taken from the same block, have also sometimes presented a remarkable difference in hardness. As these must, on account of their connection, have been nearly equally exposed to the elements, to the action of which one might at first be ready to ascribe the softening, I cannot but regard it as the consequence of a difference in the original formation, or in the composition of the stone.

1

The impatience of my guide did not permit
me to make a greater number of observations
in this gallery. The Count de Beust, who had
visited it some days before, observed the occurrence of bole; and he told me, that a column of basalt is to be seen traversed by a bed
of that substance.

9. The mountain called *Pœhlberg* is situated about forty-three miles to the S. S. W.
of Freyberg, and about twelve from the ridge
of the chain. Its name is derived from the
Pœhl, a small stream which washes its eastern
base. Another rivulet, very much confined in
its bed, bounds it on the western side. On its
western declivity is built the small but hand-
some town of Annaberg.

Pœhlberg has the form of a huge truncated cone, which rises like a colossus a-
bove all the neighbouring mountains. At a
distance, it presents a most majestic appearance.
It requires near an hour and a half, to go from
the western foot to the basaltic platform of
this mountain. The acclivity is gentle, up
to the beginning of the platform. Its
sides are covered with lands under tillage;

but there are no trees. On the top nothing is
to be found but fragments of basalt, and some
moss.

The body of the mountain is composed of
gneiss. The beds of this substance are nearly
horizontal, though in some places they have a
gentle inclination to the east. The interior is
traversed by a great number of metalliferous
veins, some of which still continue to be
worked. Towards the upper part of the
acclivity of the mountain, the gneiss is covered with gravel, which extends to the neighbourhood of the town. Over this gravel lie
several nearly horizontal beds of fine sand and of
clay ; presenting an alternation of whitish and
yellowish bands. Above these, is situated the
large basaltic platform which forms the summit of the mountain.

The shape of this platform is oblong. It is
1600 feet long, from N. to S., and 550 wide.
Its thickness is about 160 feet. Its upper surface
is nearly horizontal, being only a little inclined
to the west. It consists of a mass of basalt,
divided into irregular columns : it is difficult to
determine its structure more minutely ; for it
is almost every where covered with moss and

with fragments of basalt. I did not observe
the rock laid bare in its original position, any
where but towards the N. W.: it is there
that the prismatic division is observable.
The columns are quite like those of Scheibenberg, only perhaps they are somewhat
shorter.

The basalt is also of the same quality as that
of the last-mentioned mountain; only it appeared to me to contain a greater quantity of
grains or crystals of basaltic hornblende.

On the west side, above the village of
Annaberg, some galleries have been opened in
the sand and clay. This last substance is used
by the potters; and the sand or gravel serves
for garden-walks. The openings of these galleries, are situated merely under heaps of blocks
of basalt, detached from their native site; but
some of the galleries being more than 130 feet
long, must extend under the solid platform of basalt, which is thus proved to rest directly upon
beds of clay and of sand, at least at this part.
I myself, indeed, had no opportunity of exploring these galleries: but they were described
to me by an old miner, who had worked for thir-

ty years at them; and as he appeared to be a
plain honest man, nowise inclined to exaggerate,
I think myself justified in trusting to his information. He told me, that the roof of the large
gallery was formed by the inferior surface of
the basaltic platform; that the clay which was
immediately next to the basalt, adhered
strongly to it, and was very greasy to the
feel: that above this, the gravel was found:
that they had ceased to push the gallery forward, because they had come to a *rotten rock*.
By this expression, he meant a hard blackish clay, which I consider to be wacke. This
miner further informed me, that he remembered to have seen in the place where the basalt
is laid bare, to the N. W. of the platform, basaltic columns more numerous than they now
are; for, that, every year, some are displaced
during thunder-storms, or by the action of
frost. The same thing was affirmed to me of
the columns at Scheibenberg.

10. The *Bœrenstein* (Bearstone) is a mountain situated about six miles to the south of
Annaberg. Like the two former, it is shaped

like a truncated cone; or rather, perhaps, it
may be described generally as a lofty mountain,
having a cylindrical platform of basalt situated
on its back.

The body of the mountain is composed of
gneiss, the beds of which are nearly horizontal. As in former instances, a few metalliferous veins occur, which have given rise to
some mining labours. On the upper part of
the mountain, there is likewise a thin bed of
sand, on which the basaltic platform immediately rests. This platform is 650 feet in
length, and near 400 in width. Its thickness appeared to me greater than that of the platforms
of the two preceding mountains : I suppose it
may be 250 or 300 feet. Its sides are vertical; so that, at a distance, it seems to be much
more detached from the body of the mountain,
than the platform of Pœhlberg, from the sides
of which there proceeds an acclivity which joins
the debris of the mountain. The basalt is exposed through almost the whole extent of the
platform. It is divided into thick ill-shaped
columns of great length. On the side next
the town, there is a small acclivity, through the

turf of which, in one place, a rock penetrates
this rock is divided into small irregular columns,
from ten to fifteen inches thick, and a little inclined to the horizon.

The basalt is much of the same nature as the
preceding : it contains, however, less hornblende ; and I have found in it some grains of
iron-sand, (Eisensand).

At the foot of the mountain, on the south
side, a ditch has in former times been dug in
the sand : and I think I have observed some
vestiges of an ancient gallery, which had been
formed under the basalt.

The three mountains which have been now
described, are so situated in respect to each
other, that their summits constitute the three
points of a nearly equilateral triangle, the sides
of which are above six miles in length. They
form a striking group by reason of their singular aspect, and still more on account of the resemblance, in respect to shape, qualities and
superposition, which subsists between the
masses of basalt which respectively cover
them.

11. At the distance of twelve miles to the south of Bœrenstein, and precisely on the ridge of the chain, is situated the *Spitzberg*, (or Pointed Mountain), which, for height, vies with the Fichtelberg,—a neighbouring mountain, the summit of which, as well as the body, consists of gneiss or of mica-slate, and which forms the most elevated point of the chain; being near 4000 feet above the level of the sea, according to the barometrical ob·servations of Charpentier. The peak of Spitzberg, is composed of basalt. In shape it greatly resembles that of a mountain soon to be particularly described, (Geissingensberg, § 17). Here the basalt really rests on the mica-slate, which constitutes all that elevated part of the mountain. Its base is covered with mossy turf. In descending the rivulet, as far as Irgang, we meet continually with peat-mosses, or boggy places; and it appears, that the soil of this sort of elevated plain is derived from basalt. From Spitzberg, may be seen the Hassberg, Presnitz, and other basaltic mountains, whose summits command the whole chain, and at the bottom of which are situated the deepest valleys. As

these mountains belong to Bohemia, I do not here enter on them : those desirous of information, may consult the writings of Dr Reuss *.

12. *Heidelberg* is a small village near to Seiffen, and about four and twenty miles to the south of Freyberg. To the east of the village, there is a mountain nearly covered with fine trees : it consists of gneiss passing into mica-slate. On the declivity (not on the summit) which looks towards the town, we find two groups of basaltic columns : they have the appearance of two horns, which seem to issue from the side of the mountain. Each group is about sixty or seventy feet in circumference at the base : in both groups, the columns diverge a little. These columns are small, not more than eight or ten inches thick, and very irregu-

---

* The writings particularly alluded to, are, 1. *Orographie des Nordwestlichen Mittelgebirges in Böhmen,* 8vo. Dresden, 1790. 2. *Mineralogische Geographie von Böhmen,* 2 vols. 4to, 1793, and 1797. 3. *Sammlung naturhistorischer Aufsätze mit vorzüglicher hinsicht auf die Mineralgeschichte Böhmens,* 1796. 4. *Mineralogische und bergmännische Bemerkungen über Böhmen,* 8vo, 1801. T.

lar.   The basalt of which they are composed,
is black and compact: it contains some grains
of olivine ; and, in small cavities, a sort of marly
earth may be observed.

13. The mountain of *Lichtewalde* (Light
Forest) lies about eighteen miles and a half to
the S. S. E. of Freyberg, exactly on the frontier
of Saxony and of Bohemia, but within the ter-
ritory of the latter.   Its eastern and northern
base is surrounded by the small river called
Flœhe, which forms, in this place, the boundary
between the two States.   The mountain is situated on the ridge of the chain, and entirely
commands it in all that district.   The prospect
from its summit embraces Freyberg, and all the
anterior part of the chain, even to the plains of
Saxony.

It has the form of a great truncated cone, or
rather of a very elevated platform, which is
more than 3000 feet in diameter.   On the north
and east sides, it is isolated and separated from
the rest of the chain, by the valley in which the
Flœhe runs ; but it is joined to the chain towards
the west and the south.   Its summit alone re-

mains detached on every side. It is covered
with a pine forest, excepting on the northern
side, which is naked. In the middle, appear
the ruins of the Castle of Lichtewalde, formerly
one of the seats of Count Wallenstein, so celebrated in the history of the Thirty Years War *.

The body of this mountain, together with all
the surrounding rocks, consist of granite. The
felspar here prevails considerably : it is in large
reddish grains ; the quartz is grey ; and the mica,
which exists only in very small quantity, is
of a brownish-black colour. I found fragments of porphyry towards the upper part of
the mountain ; a fact which leads me to believe that, at least in some places, this kind
of rock covers the granite.

Above all, lies the basalt. It forms a plat
form which is above 3000 feet in diameter.
I could not determine its thickness, as it is impossible to see the surface of superposition, owing to all the sides being thickly covered with

---

* The Count was the richest nobleman in Bohemia,
and possessed several extensive palaces and castles on
his different estates. This was one of the strongest.
—For his exploits, see Schiller's History of the Thirty
Years War, &c.   T.

earth or with moss. However, as I observed granite to be the rock, 650 feet below
the summit, I can conclude with certainty
that the basalt is not of that thickness. The
same obstacle prevented me from ascertaining
whether the basalt of the platform was inclined
to prismatic division. Its upper surface is nearly
horizontal. On the top, there is a small shallow
lake ; its surface extending to about 500 square
feet, and it being little more than six feet
deep. In the court of the Castle, there is also
a well, of small depth, and which becomes dry
in the summer season.

The fragments of basalt which occur on the sides
of the mountain, are of a greyish-black colour,
uneven fracture, and very compact : some varieties which consist of granular distinct concretions, are of a deeper black. Many of the masses
on the summit, are of a large size ; some of
them from six to ten feet in height. The substance of many of them appears full of small
tortuous cavities, somewhat like those observable in certain kinds of worm-eaten timber.
These masses of basalt, are thus lighter and
less hard than the common sorts ; but in all

other respects they entirely resemble them.
I have often seen blocks, in which the one extremity presented the spongy appearance just
described, while the other was quite compact;
the cavities diminishing gradually in number
and size, till they wholly vanished.

The basalt of this mountain is principally
distinguished by the fine olivine which it contains in great profusion. It occurs in amorphous masses, sometimes larger than the fist,
and composed of granular distinct concretions.
Many of the grains of this substance, have an
asparagus-green colour ; and sometimes a very
delicate apple-green : they are shining and
transparent : the fracture is conchoidal : there
may, however, I think, be observed, a tendency
to a sort of prism, with two planes cutting
each other at right angles, which indicates a
two-fold cleavage under that angle ; and it is
certain, that the greater part of these grains
split into fragments, shaped like small rec-
tangular parallelopipeds, the bases of which are
not planes. The other grains have the appro-
priate colour of olivine : it passes, by decomposition, from green to yellow, and from that to

brown; the decomposed grains having an
earthy aspect. In some places, a part of the
substance is totally destroyed by decomposition, so that, in the space formerly occupied by
the olivine, there remains only a brown honey-
combed substance, resembling a sort of net-
work. I observed likewise in the basalt of
this mountain, some grains of a black mineral,
with a conchoidal and remarkably shining fracture, and very hard: it appears to me to be related to augite; but I shall treat more particularly of it in describing the mountain called
Heulenberg.

14. Somewhat more than six miles to the
eastward of the Lichtewalde, near the source of
the Mulda, on the frontiers of Bohemia, there
is an oblong mountain, rising to a greater
height than any which surround it. It is called *Steinkopf*. The lower part of the mountain
is composed of gneiss; the upper of porphyry.
The basis of the porphyry is red, and clayey.
Besides crystals of quartz and felspar, it contains numerous grains of hornblende: it seems
therefore to be allied to the sienite, or rather

sienite-porphyry, of Werner *. There have, in former times, been considerable ironstone mines worked in this rock. Over the highest part of the mass of porphyry, is placed a cap or small platform of basalt, forming the summit of the mountain. This cap is about 100 paces in diameter; it is covered with a grassy turf, through which some basaltic columns are seen. On the top of the basalt, a cavity presents itself, from six to ten feet deep, and about thirteen feet wide: the walls are formed by the heads of columns, which diverge like the rays of a sphere, or rather of a hemisphere, the centre of which were placed in the middle of the opening of the cavity. The basalt is very hard, and of a deep black colour. It contains grains of olivine, the fracture of which seemed to shew a tendency to the lamellar; and likewise grains of magnetic ironstone.

15. Near to the road from Dresden to Freyberg, and between twelve and thirteen miles

---

* Brochant, vol. ii. p. 576.

E. N. E. from the last-mentioned town, a mountain may be observed of greater size than any
in its neighbourhood. It is called *Landberg*
(the Country Mountain). It is situated between the village of Hertzogswalde, and the
valley which contains the village of Grund.
It is of an oblong shape : its upper part is
rounded ; and the ridge of this part runs parallel to the valley. It is covered with wood.

The fundamental rock is gneiss : on the side
next to Herzogswalde, it is covered with clay-
slate, but every where else with a porphyry
having a basis of compact felspar. Above this,
there is a bed of sandstone, on which the basalt rests. In some places, there may be observed, immediately under the basalt, a sandstone with a quartzy cement : the grains or
pebbles which are agglutinated by this cement,
are also of quartz, and they are sometimes as
large as one's fist. This sandstone, with a siliceous basis, must not be considered as a mere
variety of the sandstone which covers the porphyry, the cement of which is a ferruginous
marl : it is a distinct species, belonging to the
newest floetz-trap formation. I have observed

it likewise in the basaltic mountains of Bohemia and of Hessia.

The basalt of the Landberg covers a space
of between 1700 and 1900 square feet; but it
is nowhere exposed to a sufficient extent, to
enable me to determine its precise dimensions
in length, breadth and thickness, or to mention the kind of division to which it is inclined. Towards the top of the mountain, at the
outlet from the wood next to Herzogswalde, I
found a quarry opened in the basalt: the basalt was there divided into layers, so that it had
a schistose appearance. This division into layers, must not be mistaken for a kind of *stratification;* for the *seams of stratification* stretch
to great distances in rocks; they generally undergo different inflections, and they run nearly
parallel among themselves: whereas here, there
were only groups of small straight laminæ; in
each group, the direction of the layers was different; so that, for some time, I thought I saw
only a collection of detached masses, heaped
upon each other without any order, supposing
every group of layers to be a separate mass:
but at the bottom of the quarry, I perceived

one group, the size and position of which did
not leave any doubt in my mind that the whole
were in their original situation. The basalt of
this mountain contains some grains of olivine;
but otherwise, it presents nothing particular.

On the eastern declivity, near the middle
of the forest, and in the rock, there is a cavity,
which is called *the dog's pit*, (Hundsgrube),
and which the schoolmaster of the village, who
acted as my guide, assured me was the crater of
an ancient volcano. It is a hollow about ten
feet deep, and eight feet wide; the bottom is
covered with fragments of basalt, which have
been thrown in by passengers, and which render it impossible farther to explore the depth
of this supposed volcanic abyss. Its walls are
hung with moss; they consist of a sort of
earth, rough to the touch, of a dark grey
colour, and which has probably been produced
by the decomposition of the basaltic rock in
which the hollow is formed. The roughness mentioned, seems to proceed from small
grains of the basalt, not yet reduced by decomposition. There occur also larger pieces
of solid basalt; they are rounded, almost like

D

pebbles, although every appearance convinced
me that they were still lying in their original
position : they are merely parts of the basaltic
mass, which, perhaps in consequence of a difference in composition, have better resisted the
decomposing power. I further found in this
earth, small masses of a species of lithomarge,
bearing a general resemblance to certain hy-
drophanes : it is of a bluish-white colour; but
its surface presents some yellowish-brown spots
and rays, occasioned by a ferruginous ochre:
its fracture is dull, and small conchoidal : the
fragments are sharp-edged : it is translucent on
the edge : soft, and very soft : smooth, or
somewhat greasy to the touch : very easily
frangible : it acquires a little lustre when scrap-
ed : adheres very slightly to the tongue : placed
in water, it splits with a slight confused noise,
and falls to the bottom, without forming a
paste.

16. The small hill known by the name of
Ascher-hubel, (Hill of Ashes), is situated on the
continuation of the Landberg about a mile
and a half towards the south-east, near to some

houses called *Speisserhaus*, and in the middle
of a forest. The body of the hill consists of sandstone, and its summit of basalt.
This summit is of the shape known by the name
of *dos d'ane;* the ridge being inclined some degrees
towards the S. S. W. It is 230 feet long, and
130 broad. The most elevated point of the
ridge is not thirty feet above the sandstone.
On the S. S. E., and N. N. E. declivities, no
fragments of basalt are to be found at the distance of more than ten paces from the beginning of the basaltic summit : on the south side,
the fragments extend 150 paces down : and on
the north side, where the declivity is most rapid, I did not discover the sandstone nearer
than 300 paces ; but as the whole is covered with grass-turf, I cannot say that the basaltic rock extends to this distance. I am inclined to believe, that it does not extend much beyond the *dos d'ane* portion of the hill, the dimensions of which have been already mentioned.

A quarry which is situated on the highest
part of the basaltic eminence, enabled me to
observe the structure of the basalt. It is divid-

ed into irregular columns, which stand nearly
vertical; they are from six to ten inches thick,
and from six to ten feet high. One extremity of some of these columns consists of granular distinct concretions, while the other extremity is entirely compact. This basalt
contains some grains of olivine, the colour
of which passes from green to brown, according to the degree of decomposition it
has undergone. I also observed in it small
nodules or balls, some of them the size of a walnut, consisting of a whitish earthy matter,
opaque, and rough to the touch. And in some
places of the basalt, I likewise found fragments of sandstone, many of them about two
or three pounds weight.

17. The small town of Altenberg, situated
twenty-five miles to the S. E. of Freyberg, is
built on a great oblong mountain, called
Geissingensberg. It terminates *en dos d'ane,* and
its ridge runs parallel to that of the chain;
from which it is distant about three miles, and
which it nearly equals in height. About the
middle of the mountain, a large protuberance

appears, the aspect of which recalls the idea of the hunch on the camel's back : it is covered with fir-trees : the rest of the mountain consists of tilled lands, and of pasture-grounds.

The mountain is more than three miles long : it is traversed by a great number of subterraneous passages, intended for the working of mines ; among others, by a gallery more than a mile long. Notwithstanding these advantages, I am not able to describe with precision or confidence, all the particulars of its structure. I shall mention what appeared to me the most obvious facts. Towards the middle, its mass consists of a quartzy substance, impregnated with chlorite earth, which gives it a green colour; this substance contains tin-ore, disseminated in small grains, sometimes scarcely visible to the naked eye. This ore is the object of the great mining operations of Altenberg. Towards the west, may be observed, incumbent on this quartzy matter, a great mass of porphyry, with a basis of *hornstone,* (probably compact felspar), containing chiefly a large quantity of crystals of quartz, belonging to the dodecahedral variety described by

Haüy, and of a dirty smoke-grey colour. Towards the east, again, the quartzy substance is
covered with gneiss; but the outgoings of
the strata of gneiss do not appear at day;
they are hid by a porphyritic sienite, which constitutes the eastern part of the mountain. This
sienite is composed of felspar in large grains, of
a deep red colour, and of a little common
hornblende, in small grains, of a greenish colour: it likewise contains fine crystals of felspar,
nearly two inches long; they are of a delicate
flesh-red colour, and appear to belong to the
variety *bibinaire* of Haüy.

From the middle of the mountain, thus composed, the protuberance already mentioned
projects. It is shaped like a large flattened
dome: its circumference is more than 3000
feet, and its ascent about 160 feet. A cart-
way which conducts in a spiral direction to its
summit, and in forming which it has in some
places been necessary to cut about four feet
down, afforded me an opportunity of ascertaining that the whole of the surface of this
dome is composed of loose pieces of basalt,
which are evidently fragments of prisms; their

form proves also, that they have not been rolled; they lie in a black earth, which is the product of their own decomposition. On the
highest part of the dome, there is a group of vertical columns, which plainly remain in the same
place where they have been formed: they are
about ten inches in thickness, and their bases
form very irregular hexagons. Probably the
whole dome is composed of similar columns,
which may be hid by the fragments visible at
the surface.

The basalt of this mountain is in general of
a deep greyish-black; its fracture is small-
grained uneven: it is very compact, hard,
and difficultly frangible. It contains a great
quantity of olivine, in larger or smaller pieces,
some of them the size of hens eggs; they are composed of granular distinct concretions; in some
grains, the fracture seemed to shew a tendency
to the foliated, but in the greater part it is
conchoidal: the colour varies from apple-green,
and asparagus-green, to olive-green. Olivine,
as is well known, decomposes very readily; it
hence happens, that in all the numerous fragments of basalt which are found on the moun-

tain, the space which it occupied at their surface is entirely empty. I examined with attention the form of these spaces; it is evidently the same which the grains of olivine had assumed during their formation : the greater part
are, it is true, irregular and angular, but some
of them seemed to me to incline to a kind of
regularity : a section of one sort shewed a
hexagon, of which the two opposite sides were
longer than the others; while other sections produced a rhomboidal figure : perhaps these last
may have been made lengthwise, or perpendicular to the length of the grain of olivine, while
the former may have been made crosswise, or parallel to the grain. I do not mean to affirm,
that the olivine which occupied a similar space,
was precisely a crystal possessing the form which
would result from the combination of these two
sections; but only that some inference may perhaps thence be drawn concerning its crystalline
forms: for 1 have found, even in basalt *,
masses of felspar, the form of which approach-

* NOTE E.

ed that assumed by this mineral when its crystallization has not been constrained.—The basalt of Altenberg contains, besides, some widely scattered small grains of calcareous spar. And I observed among the olivine, a small crystal, which, from its deep colour, its fracture, and its hardness, I was inclined to consider as augite; but I did not examine it with sufficient precision to enable me to decide.

18. *Luchauerberg* is a mountain situated about eighteen miles to the W. S. W. of Freyberg, and between two and three miles to the S. W. of the small town of Dippolswalde. It is shaped like a cone, nearly isolated on every side. Its summit is about 1000 feet above its base, and forms the most elevated point in that neighbourhood. The upper part of the cone is covered with wood; the rest is occupied by pastures and cultivated fields.

The base and the chief mass of the mountain, are of gneiss; which, like all the surrounding primitive rock, is covered with porphyry. On the west side, porphyritic sienite occurs:

on the east, we find only the ordinary porphyry of the country; its cement or basis appears
to be compact felspar; it contains dodecahedral
crystals of quartz, of a deep smoke-grey colour.

A summit or cone of basalt rests upon this
porphyry; its height is about 160 feet; and
it inclines to the horizon at an angle of
20 or 30 degrees. The naked rock nowhere
appears; the whole cone being covered with
amorphous fragments of basalt, which are confusedly heaped together. The people say that
the summit is entirely composed of such fragments; nothing, however, can be more certain,
than that the centre is solid.—The basalt is of
the same general nature, as that at Altenberg;
but it contains only a few grains of olivine and
of hornblende.

When I had reached the summit of the mountain, I was overtaken by a storm accompanied
with heavy rain. I sought an asylum; and I
found one, though it proved but slender,
under the branches of a small tree, which
grew in a little hollow, between six and seven
feet deep, situated exactly at the top of the

cone. A vulcanist would not have failed to
regard this hole as an ancient crater. I descended therefore into this domain of Vulcan,
willing to avail myself of the shelter which it
seemed to offer me against the fury of Neptune. But I must confess my ingratitude ; for I
left this mountain, like those which I had already examined, without being able to re-
cognize and admire the labours of the god of
fire.

Between Altenberg and Luchauerberg, fragments of basalt are found in the fields : they are
probably the remains of some former basaltic
eminence which no longer exists. These fragments are used as a flux at the iron foundery of
Schmiedeberg. I have seen basalt employed
for the same purpose near Johann-Georgenstadt.

19. If any one, in leaving Dresden, looks
towards the chain of mountains in the S. W.,
he will perceive near the village of Cotta, a
point rising above the chain. It is called
Cottauer-Spitze, and consists of a cone of basalt, resting on sandstone. I had not personally an opportunity of examining that

mountain. I was told that the basalt is columnar; that its colour is a deep black; that it is compact; and that some traces of coal are perceptible in the sandstone.

20. Near the frontiers of Bohemia, about fifty miles to the east of Freyberg, there is a very narrow valley, which runs between high sandstone mountains, of the most picturesque appearance. The town of Schandau is situated at its outlet. In tracing up this valley, we come to the foot of the hill named *Heulenberg*. It is a huge mass of sandstone, nearly of a conical form, but joined by its sides to the rest of the chain. It forms a part of the vast deposition of sandstone which constitutes the whole rock of that country. Near the top, the grain of the rock is coarse; and its cement is a ferruginous marl, the colour of which inclines to yellow. The stratification is horizontal.

From the highest part of the mountain, two small rocky eminences project like horns. These are groups of basaltic columns, a few feet in height, and about fifty paces in circuit. The groups are about twenty-five or thirty feet distant from each other. The prisms are very re-

gular; they have six, five, and often only four
sides; and they are from three to four inches
thick. When the fragments are struck against
each other, they resound like the anvil under
the hammer.

The basalt is black, and very compact.
It contains numerous grains of ironsand;
but is particularly distinguished for abounding with a remarkable mineral, which I have
found in several basaltic rocks. Messrs Reuss,
Von Buch, and some other mineralogists, have
noticed it; but it has not yet received any ap-
propriate name. It is of a black colour, which
on the edges of small fragments seems to be
a greenish-black. It is found in grains, the
size of which varies from that of a pea, to that
of a walnut. Internally it is shining, and of a
vitreous lustre, approaching sometimes to metallic. It divides into fragments which have
the form of parallelopipeds, and the lateral faces
of which are smooth and splendent;—facts
which lead me to suppose that this mineral
may have a two-fold cleavage, the one in-
tersecting the other nearly at right angles. As
to the transverse fracture, it is perfectly small

conchoidal. The fragments are partly of an indeterminate form, and partly consist of small
parallelopipeds: they are translucent on the
edges. This mineral is besides very hard.—
From the description now given, it appears to
be intimately related to augite *.

Near to Heulenberg, there is a still higher
and more bulky mountain, called *Winterberg*,
(or the Winter-Mountain). I did not examine
it; but after my return from my visit to that
country, I learned that its summit is covered
with basalt of the same sort with that which I
have just described.

21. *Stolpen* is the most remarkable of all the
basaltic mountains of Saxony, for the beauty
and regularity of its columns. It has attracted
the attention of naturalists, ever since the days
of Agricola †. It is situated about eighteen miles
to the east of Dresden, and forty-three to the

---

* Probably only a variety of augite.   T.

† George Agricola the distinguished German mineralogist and metallurgist, in the 16th century, is here alluded to.   T.

E. N. E. of Freyberg. The small town of Stolpen is built on its declivity.

The mountain is of an oblong shape : its ridge
lies parallel with that of the chain, from which
it is distant somewhat more than twelve miles.
Towards the middle, in the long direction,
there appears a sort of swell, which assumes the
form of a cone, having its sides inclined to the
horizon at angles of from 10 to 15 degrees.
Over this, lies a mass of basalt; and upon it
is built the Castle of Stolpen, which is flanked
with four high towers; so that this spot is the
most elevated and conspicuous in that part of
the country. From the bottom of the swell, to
the surface of the court of the castle, is above
200 feet of perpendicular height; and the
summit of the mountain rises 800 feet above
the plains of Leipsig. The whole body of the
mountain consists of granite : the felspar is of
a bluish-white colour, in grains of different
sizes; the quartz is ash-grey; and the mica,
which is generally in very small quantity, is
blackish : in some places, however, this granite is very small-grained, and abounds with
mica.

The basalt seems to rest on the granite : I cannot however assert that it is immediately superincumbent; for in some places I observed, above the granite, a sort of wacke, which may prove to be interposed between them. The rock being almost every where covered, I had no opportunity of seeing the surface of superposition : it would be difficult therefore to ascertain the point where the granite stops, and the basalt begins. I can only say, that I always observed the granite to be situated at least 300 paces from the walls of the castle, on the south-west, the east, and the north sides. It seems probable, farther, that the mass of basalt commences where the acclivity becomes suddenly rapid, and indeed nearly vertical. If this be the case, the basaltic mass, as far as it is exterior to the mountain, must be about 650 feet long, 325 broad, and 130 feet high : its form must be nearly that of a truncated pyramid, the base being a parallelogram, and the sides having a very small inclination. Its upper surface is surrounded and covered by the walls of the castle. In the centre the mass is thicker, and sinks deeper in the mountain, than at its

lateral edges; for in the court of the castle, a
well has been dug to the extraordinary depth
of 290 feet: the water in this well stands 228
feet below the surface of the court, and on a level
with a small lake situated at the foot of the swell
formerly mentioned. The nature of the rock,
at the bottom of the well, is not known; but
some years ago, a mineralogist, whose accuracy
may be depended on, descended to the surface
of the water, and he declared that the walls of
the well still continued to be of basalt at that
depth. It thus appears, that the basaltic mass
is at least 228 feet thick at its axis or centre;
while at the part of its margin which I examined, it was only from 130 to 160 feet in thickness. No doubt, I cannot positively assert
that it is not thicker at some places; but that
does not seem probable. I conclude, therefore,
that the basalt here fills up a hollow, which
existed in that part of the mountain before
the basalt was deposited.

The summit consists entirely of a collection
of columns, placed beside each other like pillars in a building. The greater part are regular, and six-sided; but I observed some with

four, five, seven or eight sides. They are from
eight to eleven inches thick ; and fourteen or
sixteen feet, or perhaps more, in length. The
respective breadth of their lateral planes, varies
considerably, according to the inclination of the
one to the other; while, as I have already
remarked, the general form approaches that of
a regular hexagonal prism. To the westward,
below the castle, is situated the great quarry:
the columns are here nearly in a vertical position, only a little inclined towards the north;
they are traversed, through the whole extent of
the quarry, by parallel fissures, which divide them
into floors or ranges, placed one above another:
the workmen use no other tool, but a simple
lever ; with it they easily dislodge the columns,
which do not at all adhere together : frequently indeed they are separated by a thin bed of
earth, which is produced perhaps by the decomposition of the basalt, and which, when dried,
gives the columns the appearance of being
clothed with a sort of bark. To the south, the
rock is covered with grass. Towards the S. E.,
a basalt rock projects from the turf: it is divid-

ed into plates about four inches thick, and
which have an inclination of 60°, or 80° towards the west. To the east, on a terrace in
the garden of the overseer, there is a fine group
of vertical columns, from three to six feet
high, more or less. Towards the north, on
the margin of the castle-ditch, some scattered
columns may be observed rearing their black
heads above the grass and bushes which cover
the ground. In the court of the castle there
has been allowed to remain a fine group of
columns, rising to the height nearly of twenty
feet, and about fifteen feet in circumference. In
another part, the floor of the court consists of a
section made transversely through the prismatic
columns, which are so regular in shape,
and so nicely arranged by nature, as to present the exact appearance of a pavement made
with hexagonal flags. The walls of the castle,
and the fortifications, are built with columnar
masses, bedded on each other.

The basalt of Stolpen is black, with a bluish
tint. When a column is broken, the fracture
appears even, or very fine grained : when a

small bit is chipped off with the hammer, it is imperfectly large conchoidal; and the fragments are sharp-edged. In some places, particularly to the S. W., it occurs in granular distinct concretions, which are very easily detached. I have seen columns which were granular at one extremity, while they were perfectly compact at the other.

This Stolpen basalt, which is so hard as almost to emulate iron, contains some small round cavities, the walls of which are often invested with a coating of calcedony, thus forming a small geode. The interior is sometimes set with quartz crystals; and sometimes filled with green steatite. At other times, these cavities contain small balls of calcspar, of zeolite, and of a lithomarge resembling semiopal. The basalt further contains a few small grains of olivine, and a number of black shining points, which are either hornblende or augite: these points are hard, striking fire with the steel. In a small mass of a substance which appeared to me to be olivine, I observed a cavity studded with crystals of quartz.

The columns of Stolpen are employed as posts or boundary-stones on the roads, and in the streets of the neighbouring towns; and they are sometimes placed in gardens and pleasure-grounds, as objects of decoration. The basalt is used also in the manufacture of bottles. While walking about the church-yard of Stolpen, I remarked a group of basaltic columns, tastefully placed over a grave. Such a sight could not fail to bring to mind the recent loss of the celebrated geologist of France (Dolomieu), who has treated so well of basalt; and I could not help thinking that a similar simple mineralogical trophy, might with propriety be raised over his tomb, as calculated to perpetuate the memory of his labours. In former times, a sphere circumscribed by a cylinder was placed over the sepulchre of Archimedes,—a spiral logarithmic curve was inscribed upon the monument of Bernouilli; and both emblems were equally fitted to recall the attention of posterity to the discoveries and writings of these eminent men.

22. In some places in Saxony, basaltic sub-

stances are found constituting the mass or body of *veins*.

In going from Tharand to Dresden, rather less than three miles from the latter place, near the opening of the valley called *Plauischen-Grund*, and on the left hand, two veins may be observed in a sienite rock. They have an in-clination of about 80° to the north : the one is three feet thick, the other about one foot. Their mass consists of a substance which Wer-ner has considered as basalt *, but which appears rather to be a greenstone, or hornblende rock : its colour is a very dark greenish-grey : at first sight, its fracture seems earthy ; but when ex-amined with attention, a great number of small filaments are observable, which are nothing but common hornblende, (a substance intermediate between the *amphibole* and *actinote* of Haüy) : its fragments are blunt : in hardness and in specific gravity, it is inferior to true basalt ; and some specks of mica are discernible in it. The mass of these veins is divided, by fissures,

---

* Theorie des Filons, p. 95.—(p. 77. of Dr Ander-son's translation, 8vo. Edin. 1809).

into very irregular prisms, placed horrizontal-
ly above one another, like logs in a timber-
yard.

About thirty miles to the south of Freyberg,
near Volkenstein, the road is traversed by three
veins of a substance somewhat allied to basalt.

Not far from *Annaberg* and *Wiesenthal,* there
are still other veins of a substance having a
close analogy with basalt. The quarriers call
it *wacke;* and Werner has retained that name
for it in his Oryctognosy *. In the mine of
Marcus-Rœhling, hard by Annaberg, I saw
a vein composed of that substance : it is blackish, and of an earthy aspect; so very hard at
first, that the miners feel it difficult to push
their galleries through it; but when it has
been exposed for some time to air and moisture,
it softens and affords a sort of unctuous and
blackish clay : it contains a great number of
small white and powdery spots, on the nature
of which I cannot pronounce. In another
mine called *Gallilœische-Wirthschaft,* a short

E 4

See Note, *anteà*, p. 4.

distance from the former, veins occur, composed of wacke of a similar quality: though its
fracture is earthy in the small, it is conchoidal
in the great; that is, when large masses are
detached, the surface presents alternate concavities and convexities. At the extremity of
a gallery, which had been pushed forward into
a vein of this substance, I found it very hard;
but a few yards backwards, in a situation
where it had been acted on [by water filtering
through the gallery, it was so completely
softened, that I detached plates of it with my
fingers, which I could mould and divide like a
paste of fine clay. It contains a great number of round cavities, some empty, and others
filled with calc-spar; it likewise exhibits fine
crystals of black mica, and some grains of
hornblende. Its colour has a greenish tint;
but some pieces which I observed out of the
mine, and among the debris, were of a black-
ish-brown colour, hardened and split into
chinks. This vein of wacke accompanies another, which contains ores of silver, and which
appears to be older.

Since I have thus mentioned a vein the mass of which has an amygdaloidal structure, that is, contains small vesicles, partly empty, partly filled ; I shall state another instance, which I observed in the tin-mine of St Philip, near the town of Graupen in Bohemia. The mass of one of the veins in this mine, consists of a kind of wacke, or greenish clay, smooth and soapy to the touch, but very hard : it is full of vesicles, of the shape and size of a pea ; some of these are empty; some are lined with a layer of calc-spar, and others entirely filled with a ball of that substance. This vein cuts and traverses those in the same mine which contain tin-ores ; it is consequently newer than them.

23. In following, still towards the east, the chain of mountains of Saxony, we enter Lusatia and Bohemia. Very many of the mountains of these countries, have a covering of basalt : there are districts where that substance regularly constitutes the summit of every mountain, and where the valleys are strewed with basaltic fragments. The basalt is every where simi-

lar in nature; it contains the same foreign
matters, and presents the same peculiarities
of repository and of structure. The only
difference to be observed is, that, in Saxony,
the basaltic hills are situated near to the
highest mountain ridge: in Lusatia, on the
contrary, they are nearer to the foot of the
range. In the latter country, there are even
some which project into the plains, and stand
completely detached. I shall mention one instance of this.

Near to Gœrlitz, the chief town of a district,
there is a mountain known by the name of
Landscrone, (or the Crown of the Country).
It forms a great cone, rising to the height nearly of 1000 feet: it stands entirely isolated in
the plain, and detached from the mountain-chain, from which it is about six miles
distant. Its acclivity is gradual up to the commencement of the basaltic summit: there it
suddenly becomes very rapid. Its form is not
perfectly conical: it has rather the appearance
of two cones, nearly equal in height, which
have been inserted into each other, so that
their axes remaining constantly vertical, only

the two summits are separated. The body of
the mountain, for three-fourths of its height, is
granite, the rock of which the neighbouring
country in general is composed. The basalt of
the summit is divided into columns; sometimes also into plates or layers. On the platform above, a tower has been built: at the foot
of this, some pits have recently been excavated:
the most considerable of these is about twenty
feet deep, and as much in breadth: its walls exhibit irregular columns, or rather nearly vertical
pillars, from two to three feet thick. The basalt
is less hard and less compact than that of Stolpen. Of foreign materials, it contains only
some grains of olivine, and some shining
specks.

PART III.

RESULTS DEDUCED FROM THE PRECEDING OB-
SERVATIONS, REGARDING THE FORMATION
OF THE BASALT OF SAXONY.

§ 24. IT is a circumstance which must immediately strike an observer, in casting his eye

over the mountains of Saxony, that basalt is
to be found only on their summits; it lies *over*
all the other mineral substances of which they
are composed, but never *under* them   The interior of many of these mountains has been
well explored by means of numerous subterranean operations, undertaken with the view of
working the metallic veins which they contain; and galleries have even been carried quite
under some of the basaltic summits, (§ 8, 9,
10, 17); yet no internal cavities or openings
have been discovered, nor even any marks of
rending, which might permit the supposition
that the basalt had issued from below. The
inevitable conclusion therefore is, that it has
been deposited from above; and that it constitutes the portion of these mountains which has
been last formed.

Among the fifteen mountains with basaltic
summits, which have been described, we have
found three of the summits resting immediately upon granite, (§ 13, 21, 23); one upon
gneiss, (§ 12); one on mica-slate, (§ 11); three
on porphyry, (§ 14, 17, 18); four on sandstone,
(§ 25, 16, 19, 20); and three on beds of gravel,

sand, and clay, (§ 8, 9, 10). It seems, therefore, reasonable to conclude, that the basalt of
Saxony is of a formation less ancient than the
granite, gneiss, and other rocks which constitute the body or mass of the different mountains
of that country. Indeed it is evident, that
its formation must be comparatively recent,
since we find it to be posterior to that of loose
sand and gravel.

25. Since we find the basalt resting indis-
criminately on all the different kinds of rocks
which have been mentioned, yet possessing always the same structure and the same characters, being in fact completely and uniformly
the same substance; we may further conclude,
that in its nature it is quite independent of
the rock on which it lies, and that these have
absolutely no relation to each other.

26. When the rocks on which the basalt
rests are stratified, the surface of superposition
is not parallel to the direction of the strata:
for instance, at Scheibenberg, (§ 8), the beds
of gneiss and of mica-slate, which form the

body of the mountain, have an inclination of
60° towards the N. E., yet the surface of superposition of the basaltic platform is nearly horizontal. It hence clearly appears, that, before
the basalt had been deposited on these mountains, they had undergone considerable changes
and disintegrations; and it is therefore probable, that a great space of time had elapsed between the formation of the mass of the mountain,
and that of its basaltic summit.   (See Note A).

27. If basalt be carefully compared with
other rocks, each of its particular characters,
taken separately, may be recognized in them.—
Thus, its black colour may be observed in several species of shale, particularly in alum-slate,
in Lydian-stone, and in various rocks which
have hornblende for their basis.—Prismatical
division occasionally occurs in porphyry, marl,
burnt-clay, and in the gypsum of Mont-
Martre near Paris : Not far from the village of
Grund, between Freyberg and Dresden, there
is a mountain of porphyry, in which that sort
of division occurs in the most characteristic
manner : further, a particular variety of clay-

ironstone always assumes that form, which has led Haüy to give it the name of *bacillaire* * — There is nothing particular in the specific gravity of basalt: it is three times greater than that of water; but, not to speak of metals and their ores, we know, that in the class of stones, all those of the barytic genus, with garnets, and others, have a specific gravity more than four times greater than water.—Basalt contains crystals and grains of foreign substances; but such are found also in every rock which has a porphyritic structure. Through such rocks, the crystals are nearly uniformly and equally disseminated ; and they are interspersed in the same way in the basaltic paste, and scarcely ever grouped.—The minerals which are more particularly found in a crystallized state in basalt, and which seem indeed peculiarly to belong to it, are,

---

* Fer oxydé rouge bacillaire,—Columnar clay-ironstone of Jameson. A bed of this rare mineral occurs a little way south from the Cock of Arran ; and here " the action of the sea has exhibited a beautiful display of the regularity of its columnar concretions."—Rev. Mr Fleming's Sketch of a Tour to Arran, Scots Magazine for February 1808. T.

basaltic hornblende, olivine, augite, leucite, foliated and cubic zeolite, and magnetic iron-stone. Now, basaltic hornblende occurs in the wacke which occupies veins in Saxony: near Braunau in Bohemia, I observed olivine in a hornblende rock: augite is found in the granite of Norway: Dolomieu * mentions his having observed leucite in the rocks of the Pyrenees, and in vein-masses from the gold-mines of Mexico: I noticed a great deal of foliated zeolite in the veins of Andreasberg in the Hartz, and it is equally common in the mines of Allemont: lastly, crystals of magnetic ironstone abound in chlorite rocks.—The tesselated and porous aspect which has long been considered as exclusively characteristic of basalt, and of the products of fusion, I have remarked in quartz, in calcareous rocks, in ores of manganese, and particularly in bog iron-ores. Some of these last, which are forming almost under our eyes, in the swamps of Lusatia and Silesia, abound so much with vesi-

* Journal des Mines, NO. 27.

cular cavities, that they resemble a sponge. Such rounded cavities or vesicles, whether empty, or filled with green-earth, calc-spar, steatite, calcedony, or forming geodes, occur in many other mineral substances, particularly in the *toadstone* of Derbyshire: There is at Finkenhübel, in the county of Glatz, a rock containing petrifactions *, and at the same time abounding with such balls of green-earth

---

* M. Von Buch, in his Mineralogical Description of the Environs of Landeck, thus notices the fact : " It is very remarkable, that, imbedded among the rounded masses in the substance of this amygdaloidal wacke, some *turbinites* make their appearance; they have the same colour as the wacke; they are hollow, and in a good state of preservation; indeed they have tended to save the rock which contains them, from disintegration. If pains were taken to search other parts of the hill, more of these petrifactions might probably be found." Though I myself, when at Finkenhübel, did not happen to observe these petri- factions *in situ*, the fact of their occurrence is not the less certain; and indeed M. Von Buch sent specimens of the wacke, including turbinites, to the Museum of Natural History at Berlin. See *Journal de Physique*, t. 47. p. 157. A.

and calcedony: and I have already, at § 22.,
spoken of the mass or body of several veins, in
which I observed numbers of these round vesicles,
both empty and filled with calcareous spar.—
Basalt almost invariably occurs on the summits
of mountains ; but in Saxony, porphyry is very
often found in similar situations: I shall instance only the mountain on which is built the
fort and town of Augustus, (Augusteburg), and
the Burgberg, six miles to the S. E. of Freyberg.—Basalt often forms isolated hills of a
conical shape: In Lower Silesia, however, between Landshut and Waldenburg, the country
is covered with large isolated and conical hills
composed of porphyry: the famous sandstone
hills of Kœnigsstein and Lilienstein, about twenty miles S. E. from Dresden, likewise stand detached, and are shaped like truncated cones.—
But I shall not pursue the comparison farther.

It is no doubt true, that basalt possesses
some characteristics, some properties which
seem peculiarly to belong to it; but every kind
of rock is in the same predicament; granites,
shists, and porphyries, have all of them their
particular characters, or their habits, if I may so

express myself. It seems sufficient for me
to have pointed out in other rocks, all the characters and peculiarities presented by basalt,
to entitle me to conclude, that it may have
been formed after the same manner, or may
have had a similar origin; and indeed I am au-
thorised to believe so, because it has not been
shewn that it has had a different origin. Where-
fore should we suppose Nature to deviate from
her ordinary course, while every thing testifies
that she is not given to variableness? If an
agent or mean be extraordinary, it appears to
me that we ought not to admit it on that
very account, if there be no proof of its necessity. When indeed such mean shall be
shewn at least to be possible, which is far from
being the case here, it may then be proper to
examine it by its consequences. All naturalists regard granite, porphyry, shistus, &c., as
precipitates from a solution which had once covered the countries where they are now found :
—I therefore consider the basalt of Saxony as
a precipitate or sediment proceeding from a solution which had at one time stood suspended
over the whole of that country.

28. In Saxony, basalt occurs only on the summits of mountains. Now, its being confined to those elevated spots, can only be accounted for in one of two ways; either the basaltic sediment must have been deposited only on the tops of those mountains where we now find it; or it must, in the form of a bed or great layer, have once covered the whole country, and the intermediate portions, which are now wanting, must, from some cause, have been destroyed and removed. There is absolutely nothing to induce us to adopt the first of these two opinions : it is too improbable. Existing appearances are in favour of the second. Very narrow valleys, or ravines, generally occur at the foot of mountains which have basaltic summits : these ravines separate the mountains from one another; and are regarded, in the eyes of almost all naturalists, merely as deep furrows formed in the mass of mountain-chains. The isolated situation, and the abrupt declivities of hills with basaltic summits, are certainly not the result of their original formation; but have proceeded from the destruction of the surrounding rocks. The disintegration of former rocks is

demonstrated by the existence of vast beds of
sandstone, and puddingstone, and by the extensive sands which cover so great a portion of
the surface of the globe; for these substances
are evidently nothing but the detritus of previous rocks.   In the mountains of Saxony, the
basaltic deposition covers the other rocks; so
that these could not be affected till after the destruction or removal of the former.   A single
glance of one part of this chain of mountains, leaves no room to doubt that the basaltic platforms, visible on many of them, did
once form a continuous mass.   Let what I have
said concerning the mountains of Scheibenberg,
Pœhlberg, and Bœrenstein, (§ 8, 9, 10), be called to remembrance: They are three huge
truncated cones, isolated, and placed so as to
form the points of a nearly equilateral triangle,
the sides of which are about six miles long.
The upper massive parts of these mountains, are the highest in the neighbourhood,
and they are nearly on a level with each other.
On each of them reposes a great platform of
basalt, nearly 200 feet thick.   The basalt lies
immediately on a bed of clay; which again

rests on a fine sand, the grain of which grows
gradually coarser downwards, so that at last it
becomes gravel : the clay and the sand form horizontal beds, which, taken together, are only
some yards in thickness. The basalt of all the
three platforms, is divided into columns of the
same shape, and size; and it is absolutely of
the same nature, containing an extraordinary
quantity of small crystals or grains of basaltic
hornblende. Thus, in these three platforms, we
find the same form, the same substance, and
above all, the same circumstances attending
a very singular superposition : the identity
is so striking at first sight, that it scarcely
leaves time for reflection; the observer being
instantly constrained to conclude, that these
three platforms are parts of one and the same
whole, and that they are the remains of a continuous bed of basalt, which formerly existed
at that level. The resemblance is not, it is
true, equally striking when we compare many
other basaltic summits ; but still it is sufficient
to entitle us to draw the same general conclusion. It is perfectly well known, that a mineral bed, when of very great extent, never pre-

serves an absolute homogeneousness of substance; never contains throughout the same
foreign ingredients in similar quantity; and
never retains all the same peculiarities of repository, especially when it partly rests on
alluvial soil.—From all that has been said,
therefore, I infer, that the basaltic summits and
platforms of the mountains of Saxony, are
merely the shreds and remains of a vast deposition of basalt, which had at one time covered the whole of that country.

29. We have just seen an instance of clay accompanying basalt, and serving as its support.
This fact, which occurs in a number of places,
has forcibly struck many geologists: they have
examined into it with care, and have perceived
that basalt has important geognostic relations
with clay, or rather with an intermediate
substance which in Saxony is called *wacke*.
Werner, speaking of Scheibenberg, says, " I
here observed, in a progressive suite of gradations, the most evident transition from
clay to wacke, and from this to basalt:
these three substances, are the product of the
same formation; that is, they are precipitates

F 4

or sediments from the same solution, which, as it became more and more tranquil, had successively deposited, first the clay, then the wacke, and last of all the basalt." Dr Reuss, whose name is well known in geology, informs me, that he observed, in Bohemia, basalt resting immediately on an argillaceous substance; and that the prismatic division of the basalt, was continued into the clay or wacke; so that each column was composed of a very hard basalt at the one extremity, and of a soft clay at the other; the two substances passing so gradually into each other, that the Doctor found it impossible to assign the line of separation. This fact is related in Dr Reuss's History of Bohemia. Werner describes a similar appearance, (Note D.) I have already mentioned my having seen blocks of basalt, very hard at one extremity, and nearly soft at the other, and the probability that this state was the result of the original structure, rather than a softening or alteration which had affected only one portion of the block. All these things strongly incline me to admit the conclusion of Werner; and, in that view, basalt may be

regarded as an argillaceous mass, of a black colour, impregnated with ferruginous matter, and
whose particles have contracted a strong ad-
hesion together *.

30. When certain kinds of basalt are examined with attention, some small folia are
observable in their mass, which appear to be
common hornblende. In others, particular
parts of the mass seem to consist of an assemblage of small points, some blackish or
greenish, and others white : when these points
increase in size, it is distinctly perceived that
they are grains of hornblende and felspar with
a lamellar structure, which, by their aggregation, compose the rock ; the hornblende pre-
dominating. Almost all the *whinstones* of English authors, are of this sort. Werner has given
the name of *grünstein* (greenstone) to this granular rock composed of hornblende and felspar ; and in default of a French name, I have
adopted the German. When the grains diminish
so much in size, that the eye is unable to discern them, and the mass seems homogeneous,
the rock becomes basalt : so that the relation

* NOTE (*f*).

between basalt and greenstone, resembles that which subsists between common compact limestone, and granular limestone, *(salin.)*

I might produce a number of examples to shew the existence of this relation between basalt and greenstone, and the transition from the one to the other; but I shall mention only one prominent instance, which occurs in the most interesting of basaltic mountains which I have ever seen, *Meisner* in Hessia. This mountain rises like a Colossus above all those which surround it. On its summit is a plain or flat about nine miles in length, and three in width. The mass of the mountain consists of limestone containing shells: over this are some thin beds of sandstone, and of sand: then a vast bed of coal, the thickness of which, in some places, is ninety feet: immediately over the coal, lies a basaltic platform, more than 300 feet thick, and which forms the plain above mentioned.— (For fuller details, see Note C).

Greenstone is found almost everywhere on the upper part of the platform. In some places it has the appearance of a beautiful granite : The grains of hornblende are black or green,

lamellar, and as large as peas; those of felspar
are whitish. On the lower part of the platform,
towards the west, we find basalt with prismatical divisions, and as black, compact, and homogeneous, to appearance, as any where to be met
with. I selected a suite of about a dozen of
specimens, which presented a decreasing series,
in regard to the coarseness of the grain, from
the beautiful greenstone to the compact basalt
which have been mentioned: And lest it might
be objected to me, that these specimens may
not have belonged to one and the same continuous mass, I procured some, in which the
granular substance, but very fine-granular, is situated in the midst of the compact, and each
insensibly passes into the other.

To prove still further the relation between
basalt and greenstone, and the transition of
each into the other, I shall here produce a
testimony nowise liable to suspicion, of
high authority, and which ought doubtless to
have great weight. Dolomieu, in treating
of the monuments of antiquity which had been
brought from Ethiopia to Rome, observes, " I
have seen many statues, vases, and sarco-

phagi, formed of black stones, possessing all
the characters ascribed to basaltes, and which
have retained that name; and I can affirm with
confidence, that these stones are not of volcanic origin. Some consist of schorl in mass *, of a
foliated fracture like hornblende; but the most
common of these black stones, are of a compound structure,—a sort of granite, in which
the scaly schorl prevails so much, that the entire mass appears black: it is associated
with white felspar, the grains of which are
so small, or so interwoven with the scales
of schorl, that it is often difficult to perceive
them. Sometimes the felspar itself appears
black, while it is really transparent, and only
transmits the colour of the schorl with which it
is connected, and the hardness of which it
tends to increase. It sometimes happens that
the felspar increases in quantity, and then the
rock assumes, in that part, the appearance of a
true granite," but which is only greenstone.
" From thence proceed the veins and spots of
granite, which are found in almost all the

---

* Dolomieu means basaltic hornblende.   A.

black rock-masses called basaltes, and the explanation of which has much embarrassed those naturalists who maintain the ignigenous origin of these rocks. In examining the antique basaltes, I have traced the passage from an almost homogeneous mass of schorl [basalt], to granite [greenstone], composed nearly of equal quantities of white felspar and of schorl; the gradual transition depending entirely on the proportion of felspar and the size of the grains. These facts leave no room to doubt that all these rocks belong to the same system of mountains *." I believe it is almost impossible to say any thing more plain, or more convincing, to shew the transition from basalt to greenstone; or to draw conclusions more satisfactory concerning their identity, and their formation †.

---

* *Journal de Physique*, t. 37. p. 195.

† The relation between certain sorts of basalt and greenstone, has not escaped M. Desmarets. This author examined the antique basalt in Italy, and found it to consist of a collection of small blackish folia, variously grouped. He observed great quarries, he tells us, of a rock perfectly similar, in the Limosin near to

I shall, however, still bring forward some arguments furnished by chemistry *.

All the proofs which chemical analysis can afford to shew the identity of two substances, here concur. Two fragments, one of basalt, and the other of greenstone, analyzed by that distinguished chemist Dr Kennedy, according to the same process, afforded the same constituent parts, and, with a very slight difference, in the same proportions ; as may be seen in the following table :

Constituent Parts.	Basalt.	Greenstone.
Silica,	46	46
Alumina,	16	19
Lime,	9	8
Oxide of iron,	16	17
Water and volatile matters,	5	4
Soda, about	4	$3\frac{1}{2}$
Muriatic acid, about	1	1
Loss,	3	$1\frac{1}{2}$
	100	100

Tulle, and also in Auvergne. He proposes to give the name of *gabbro* to this substance. The *gabbro*, therefore, of M. Desmarets, is to be considered as a rock entirely composed of small grains of hornblende. He mentions a variety, in which the hornblende is mixed with felspar, in spots or folia differing in size and in frequency.—*Mém. sur le Basalte*, part. 3. art. i.   A.

* NOTE *(g)*.

The basalt thus analyzed, was a fragment of the famous columns of the island of Staffa; it was of a bluish-black colour, of a fine grain, and homogeneous texture; its specific gravity $= 2.87$. The greenstone analyzed, is called by the Doctor *whinstone of Salisbury* *: it was an aggregate of hornblende, and of felspar more fusible than common felspar; its specific gravity $= 2.80$ †. These two minerals, therefore, must be considered as the same substance, since we find them to be composed of the same ingredients. The only difference seems to be that, in the case of the greenstone, the constituent parts have had time to attract each other, and to coalesce according to the laws of their chemical affinities; while, in the basalt, they had united in a confused manner.

From melted basalt, a blackish glass is obtained : I have already stated, that in Saxony bottles are made of it. Greenstone affords a glass of the same quality, of which buttons are

---

* Salisbury Craig or Crag, in the immediate vicinity of Edinburgh, is alluded to. T.

† *Annales de Chimie,* ventose an 10.

manufactured in Franconia *. At Edinburgh,
Sir James Hall made experiments on the melting of *whinstone* (greenstone), and of basalt,
and obtained from both a glass of the same nature, which, on being again melted, and cooled
slowly, afforded a substance of a stony aspect,
and bearing a resemblance to basalt. In these
experiments, the heat had merely overcome the
original mode of union between the constituent
parts, whether of greenstone or of basalt; and
the slow cooling had produced a second mode of
union, which was the same in both substances.
Now, in that second state, the substances
being identical, it is fair to conclude, that they
are composed of the same constituent parts:
but they must also originally have been so, since
nothing has either been added or taken away.
In short, they naturally differ only in this, that
the union of the constituent parts is *crystalline* in
greenstone, and in basalt *confused*. So that, if
the crystalline aggregation be destroyed, and in
place of it be substituted the confused, which

---

* *Journal de Physique*, germinal an 7.

has prevailed in the formation of basalt; then the same constituent parts, which had formed a greenstone, afford a mass of basalt, as is proved by the slow cooling in the Edinburgh experiments *.

From all that I have now said, I draw this conclusion, That basalt is a compound consisting of the constituent parts of hornblende and of felspar; but that the precipitate from which it has resulted, having been rapid, or perhaps disturbed by the intermixture of some heterogeneous substances, and possibly by the superabundance of one of the principles; these constituent parts have coalesced in a confused manner : had circumstances favoured their uniting together agreeably to the laws of their reciprocal affinities, they would have formed a compound of lamellar grains or hornblende and felspar, that is, a *greenstone*.

31. This fact appearing to me to be placed beyond doubt by observation, and by the re-

---

* Note H.

6

sults of chemical experiments, and farther, to be
explanatory of the formation of basalt, and to be
essential to it; I think myself entitled to make
use of it as a definition : thus,—Basalt is a compound of hornblende and felspar, but in such
minute particles, that the eye cannot distinguish
them; they are indeed so intimately blended
together, that they produce a homogeneous
mass; in short, basalt is a compact *greenstone*,
or compact hornblende-rock.

32. I shall here concentrate the conclusions
which result from my observations on the subject of the formation of the basalt of Saxony.
It appears, that long after the formation of the
other rocks of the same chain, there has been
deposited over these rocks, a vast bed of basalt,
proceeding from a solution which had covered
the whole country; that a part of this basaltic
deposition has been worn away or destroyed;
and that the platforms and mountain-caps of basalt, which now exists on those mountains, are
the shreds and remains of that deposition.—
I believe I may add, that if the precipitate
from which basalt has resulted, had been ac-

complished slowly and calmly, and in circumstances favourable to crystallization, the same
constituent parts would have formed *greenstone:*
and that perhaps in those places, and at those
times, where the solution had been more agitated and coarser, the sediment has produced a
wacke; that is, a basalt approaching to the
nature of clays.

This conclusion is not an hypothesis; it is a
natural consequence of the facts which I have
detailed. And in the additional observations
which follow this Memoir, I shall shew how satisfactorily this doctrine affords an explanation of
the properties of basalt, and of the peculiarities
of its repository.

In order not to interrupt the connection of the
argument here, I shall place in a note, (Note I.),
the answer to some objections which may possibly be offered, and likewise some particular
illustrations; and as many readers would probably like to see the opinion of the distinguished WERNER, on the subject of the formation of
basalt, I have stated it in Note L.

—————

PART IV.

PROOFS THAT THE BASALTIC ROCKS OF SAXONY
ARE NOT OF VOLCANIC ORIGIN.

§ 33. AFTER having, on the subject of the
formation of the basaltic rocks of Saxony, established opinions which are immediate consequences from facts observed, and after having
traced in these mineral masses, the characteristic marks of the same mode of formation which naturalists have acknowledged in
all others, I might perhaps reasonably be excused from proceeding to prove that they have
not had a different origin.   But several observers, and, among others, MM. de Weltheim,
Fichtel, Ferber, De Born, Voigt, and De Luc,
having regarded these rocks as the productions of fire, and as the offspring of volcanic
eruptions, I think it necessary to shew that this
cannot have been the case.   There seems here
to be a choice of two suppositions; either that

each of the basaltic mountains of Saxony has been a separate volcano; or, that all the basaltic summits in that country, are merely the remains of one great stream of lava, which had spread over the entire chain of mountains. I shall first examine the former supposition.

34. (1). A volcanic mountain is, and must of necessity be, a confused mass of fragments of stones, blocks of rocks, rapilli*, pumice, cinders, scoriæ, and the remains of streams of lava; and all these matters heaped together in utter disorder. The basaltic mountains of Saxony, however, are constituted of granite, mica-slate, sandstone, and other rocks, altogether similar to those found in the surrounding country; both are of the same nature, and have the same structure. It is only on the

---

* *Rapilli (Lapilli)* are fragments of pumice and of reddish and blackish vesicular lavas, from the size of a pea, to that of a hen's egg, which are generally thrown out of volcanoes, along with volcanic ashes, after the eruption of the lava. They occur on Vesuvius, Santorini, Etna, and other active volcanoes: also in the extinct volcanoes of Auvergne and other countries.   T.

summit that basalt is found, and it there forms
a continuous and homogeneous mass.

(2). In the axis of every true volcanic mountain, an opening must be visible, from which
the materials composing it have issued. Many
of the mountains of Saxony, however, have
been penetrated in every direction, by mines
and levels, which have been even pushed in
some instances under the basaltic summits;
yet there has every where been found a solid
and continuous mass, an ordinary rock, only traversed by some metallic veins and strings.   In
no place has a volcanic abyss been discovered,
nor any appearance of disruption.

(3). The top of a true volcanic mountain,
presents a funnel-shaped opening, or crater;
and even when, by the lapse of time, the opening has been choked up, still a distinct hollow will remain.   The soil of this hollow will
be found to be a mere collection of loose
fragments, from the debris of the volcanic
mountain itself.   In the mountains of Saxony,
however, in place of finding a hollow on the
top, we have an elevation.   The summits of
many of them are perfectly conical; the middle

point is the most elevated, and it is the point of
a solid mass of rock.   When the cone is truncated, its superior base forms a platform, composed of a continuous and very solid rock.

(4). Even in the case of those mountains
where the subterranean operations have not proceeded under the basaltic cap, there is not
wanting evidence of the physical impossibility
that the black masses composing their summits
should have issued from the interior of the
earth.   For the expansive force supposed to
have expelled the liquid matter of the basalt,
or supposed lava, certainly would not have
pierced the surface at the very axis of a cone,
precisely at the spot which presented the greatest resistance.   It would rather have been in
a low level, towards the foot of the mountain, that the lava would have burst out.   It
is well known, that in a homogeneous surface,
the shortest line is that which offers least resistance; and though chance or some local circumstance might, in one place have produced
an eruption in the axis of a cone, it is not to
be believed that the same chance, and the
same local circumstances, would occur in all

the basaltic mountains of Saxony; yet in all of them, basalt is found only on the summits.

Some vulcanists, feeling the force of such reasoning, have alleged, that the mountains themselves, of granite, mica-slate, &c., on which the masses of basalt rest, had been elevated by the same force which had produced the eruptions. In answer, I shall merely observe, that the stratification of these mountains (§ 8.) being parallel to that of the surrounding country, is a fact which bears convincing testimony against such a supposition. Besides, such an impulse might have broken and dispersed matter of so friable a nature as that which composes granite, and other rocks, but it could not have heaved it up without derangement. Further, the supposed force must have been sufficient, in this instance, to elevate at once an entire chain of mountains; for the most of those mountains which have basaltic summits, are joined by their sides to those which stand next to them. I have already mentioned, that there is, at Altenberg, a large oblong mountain, very narrow, but above six miles long, on which there occurs a basaltic excrescence, recalling the idea of the bunch on the camel's back. A mountain of this oblong shape, never could owe its origin to a heaving up, produced by an impulse which had pierced the surface immediately below the basaltic eminence. It may be remarked, too, that the body of the Altenberg mountain, is traversed by vast galleries and subterranean excavations, which have afforded an opportunity of ascertaining that the beds of rock present no appearance of rupture or derangement.

There is so little resemblance or relation between Vesuvius or Etna, and the basaltic mountains of Saxony, that I must suppose those who have taken the latter for extinguished volcanoes, to have stood some leagues off when they conceived that idea. Mountains composed of porphyry, or even of sandstone, if they had happened to have a similar form, might equally have given rise to such an illusion.

35. Let us now consider, whether it be actually within the verge of possibility, that all the basaltic summits of Saxony can be the

remains of one great stream of lava, thrown out by a single volcano. Here I must observe,

(1). That whoever can even suppose the existence of such a volcano, or pretend to fix on its position, must absolutely know nothing of the country. In this chain of mountains, are to be found the most extensive and the most numerous subterranean galleries and levels in all Germany : for more than six centuries, have numbers of miners and mineralogists been traversing the bowels of these mountains in every direction; so that I venture to say that the interior is nearly as well known as the exterior. (Note M.)

(2). In the whole of the large district in question, neither scoriæ, nor cinders, nor scorched stones, are to be found ; nor in short, does any thing occur which bears the least mark of the action of fire, or which could excite such a notion in the mind of an accurate observer. On this subject, I may appeal to M. Duhamel, who visited the country, and is well acquainted with it.

(3). Throughout the whole chain of mountains, every summit, every platform, presents

a continuous mass, which indicates a simultaneous formation at one particular period. We
no where find here, an alternation of basalt
with other rocks, which might have permitted
us to suppose, that the basaltic formation had
been repeated at different eras, or that there
had been various eruptions of lava. A single
eruption, therefore, must have produced a torrent of lava, sufficient to cover an entire chain
of mountains, more than 100 miles long, and
from 40 to 50 miles broad. A volcano must
be supposed to have suddenly broke out, and
to have vomited forth the immense stream of
lava alluded to, and then, as if exhausted by so
great an effort, to have become extinct, and dis-
appeared so completely, that the very vestiges
of it cannot be traced! Such suppositions are
too much at variance with the facts presented
by actual volcanoes, and outrage probability
too much, to be admitted, especially without
a shadow of even presumptive evidence.

(4). The very form of the bed, is sufficient
to disprove the supposition: for, according to
that hypothesis, the lava must have overwhelm-
ed the entire mountain-chain, the ridge and

two declivities, in all their extent. Its surface
must thus have extended over more than 600
square miles, yet the thickness of the bed no
where exceeds 300 feet. All the remnants
of this bed, even those which rest on primitive
rock, are more or less elevated in situation, and
that without any order or regularity: for example, the basaltic summit of Ascherhubel
(§ 16) is on a lower level than that of Landberg (§ 17), although the former is situated
nearest to the ridge of the chain. It thus appears, that when the supposed bed of lava was
entire, it must have presented a number of bendings, or a succession of eminences and hollows;
in short, it seems to have followed all the ine-
qualities of the surface on which it was deposited.
Now, such an appearance, is certainly not that
which a stream of liquid matter would have
assumed. Lava, issuing from a crater, never
would have spread itself over a chain of mountains, so as to cover the whole ridge, and both
of its sides, with a continuous sheet of nearly
equal thickness throughout; in its progress, it
certainly would not have descended and ascend-
ed, so as to produce such an effect. A stream

of lava moves forward only in consequence of its gravity, which inclines it constantly to descend ; and it is altogether impossible, especially as its motion is very slow, to imagine that, in its progress, it should leave the low grounds, and ascend heights. Whenever lava has once descended to a certain level, we cannot, in following its course, ever find it at a more elevated point. In short, it is regulated by the invariable laws of hydrostatics : any thing alleged to the contrary, can have no real foundation ; it would imply an inversion of the laws of nature, and must be rejected by every sound mind.

It appears, then, from these observations, that it is not more possible to conceive, that the numerous basaltic summits of Saxony, are the products of a single volcano, than that they are the products of many.

ORIGIN OF BASALT IN GENERAL.

36. I now proceed to examine, whether the nature of basalt, and of the substances found imbedded in it, and the peculiarities of its struc-

ture and repository, will permit it to be considered as a volcanic product. For the sake of
greater latitude of illustration, I shall here treat
of basalt in general, and not confine myself to
that of Saxony in particular.

I again repeat the caveat to be found at paragraph 5.—I absolutely exclude every hypothetical assumption.

I regard the heat of lava as of sufficient intensity to melt stony matters; for I have no reason to believe the contrary, and lavas themselves are nothing but stony matters melted.
Some persons, it is true, have supposed, that the
substance of lava exists in a fluid state in the
centre of the globe, or below the granitic crust.
I shall afterwards take occasion to canvas this
opinion. Here I deem it sufficient to observe,
that it is a supposition purely gratuitous, and destitute of any thing like evidence. From the
descriptions of Vesuvius, Etna, and other volcanic mountains, it appears, that the fires of
these volcanoes are situated in rocks similar
to those with which we are acquainted. Among
the ejected matters, we find granite, mica-
slate, limestone, &c.; and the lavas proceed

probably from the fusion of the walls of the volcanic caverns.

Proofs derived from the nature of Basalt.

37. We have already remarked, (§ 29.), that basalt is intimately related to argillaceous wacke; and made mention of columns, in which one extremity is a true basalt, while the other is a clayey substance. The clay and the basalt, must, on the supposition of a volcanic origin, have evidently been the product of a single jet; which is impossible. This fact, therefore, is of itself conclusive against the notion of a volcanic origin. The argillaceous wacke cannot be considered as proceeding from a muddy eruption, as some have suggested; for it passes into basalt by gradations so imperceptible that it is impossible to assign any line of separation. It is equally vain to allege, that the wacke is merely a decomposed lava; for, as at Scheibenberg for example, we find wacke passing to ordinary clay, and the clay degenerating to sand, and at length to gravel; while a lava, in decomposing, could never produce a quartzy gravel.

The circumstance that no olivine nor augite
are found in wacke which lies immediately under basalt containing these minerals, is a proof
that they are two distinct rocks, and that the
wacke is not merely decomposed basalt.—For
this remark, I am indebted to M. Ramond, in
his report.

38. It has been already shewn, (§ 30), that
basalt is only a modification of greenstone. At
the Meisner Mountain, and in other places,
these substances are found so intermixed together, and passing so evidently into each other,
that it is impossible to refuse to admit that
they have had the same origin. Now, greenstone, far from being a product of fire, cannot
ever have been acted upon by that element:
for it is a rock with a granular structure, composed of grains of hornblende, which retain all
their lustre, and of grains of felspar, possessing
all their brilliancy, translucency, and freshness.
According to Saussure *, felspar melts at 70 of
Wedgwood's pyrometer, basalt at 76°, and

* _Journal de Physique,_ an 2.

hornblende at about 100°. A less heat deprives felspar of its lustre: and from the experiments of Sir James Hall, (Note N), it appears that the felspar existing in greenstone, is more fusible than common felspar. If, however, basalt be regarded as a volcanic product, it is indispensable to assign the same origin to greenstone. Dolomieu, who perceived the necessity of this consequence, has preferred considering those basaltic rocks which contain portions of greenstone, as not having been the products of fire. His words are: " The veins and large granular patches (of greenstone), which are to be seen in almost all the black rocks called basaltes, have greatly puzzled those naturalists who hold that these stones have been the product of fire."—" I affirm with confidence, that such rocks cannot have had a volcanic origin."

M. Desmarets is of the same opinion. He thus expresses himself: " Naturalists who have not seen those mixtures of crystals, or collections of laminæ of *gabbro* (hornblende) occur-

H

ring in lavas *, except in cabinets or collections, may have been led to suppose that they were the result of fire; and may have been followed by many who adopt the opinions of others, without examining for themselves. But every attentive observer will abandon such suppositions, when he is shewn the hornblende existing in the midst of substances, the arrangement and disposition of which cannot possibly be attributed to fire." He adds in another place, that " there is not to be found in compact lavas, a single vestige of those crystals of hornblende, if the unaltered part of the original rock in which the volcano is situated, do not contain them †."

39. When specimens of basalt from different countries, and even from different quarters of the world, are compared together, their general resemblance is surprising. All the specimens which I have had an opportunity of see-

---

* I believe that M. Desmarets includes several sorts of basalt under the name of lava.   A.

† *Mémoire sur les basaltes, 3me partie*, art. iii.

ing,—brought from Sweden, Hessia, Saxony, Silesia, Bohemia, Hungary, the Auvergne, Italy, and the Isle of Bourbon,—have the same black colour, the same fine-granular or compact and earthy fracture, the same specific gravity of 3.000, and the same degree of hardness; they incline to the same prismatic or columnar division; and they contain the same imbedded minerals, particularly basaltic hornblende and olivine. The chemical analyses which Bergmann, Klaproth, and Kennedy have given of the basalts of Sweden, of Bohemia, and of the Isle of Staffa in Scotland, agree more nearly together, than different analyses of the same mineral often do. This identity is important, being one of the striking attributes of rocks produced in the humid way : calcareous and micaceous rocks are every where similar ; but the lavas of Vesuvius, of Lipari, and of the Solfalare *, do not resemble each other ; on the contrary, they present very marked differences : and, at the same time, these places, it will be

* *Journal de Physique,* Germ, an 7,

observed, are situated but a short way from each other; indeed, I venture to say, that signal variations as to quality may sometimes be observed among the successive lavas of the same volcano.

If basalts really were lavas, their nature would certainly participate of that of the rock in which the volcanic fire happened to be situated. Every person has observed how very different, in quality and appearance, the rocks are in one place, from what they are in another: and from all that we can learn, from the greatest depths to which miners have penetrated, it appears that, in the interior, the change of rock is nearly the same as takes place at the surface. If volcanic fire exist in one country where there is nothing but granite, and in another, at the distance of more than 600 miles, where the only rock is clay-slate, it may surely be presumed, that the fire will not in both cases be situated in the same species of rock : the melted granite, and the melted clay-slate, would not probably each afford the very same kind of stony substance ; and doubtless *all* the lavas proceeding from granite, would not be identical with *all* those proceeding from clay-slate.   But I affirm, that there is absolutely no difference between the basalt that rests on the granite of Lusatia, and that situated on the clay-slate on the other side of the Rhine, behind Neuwied.   In both these places, and every where else, the basalt possesses the same characters.   The basalt which lies on the sandstone of the Misnia, is identical with that which rests on the shelly limestone of Hessia.   In all these countries, no other substance has ever been taken for a lava.   The nature or character of the basalt, having therefore no dependence whatever on the quality of the rocks on which it reposes ; it appears altogether unreasonable to view it as consisting merely of those rocks, altered and fused by volcanic agents.

40. Almost all basalts contain about 20 *per cent.* of iron.   Now, I would ask, Which of all the rocks that we are acquainted with, is able constantly to afford this proportion of iron ? Not one ; if we exclude basalt itself, and *its modifications*, from being numbered as rocks.   In a country where no rock but granite appears, and

where all that miners and mineralogists have
been able to ascertain concerning the order of
superposition of rocks in general, necessarily
leads us to conclude that this rock extends to
a great depth; in such a country, I say, from
whence can the basalt there found, have derived its 20 *per cent.* of iron? Felspar scarcely
contains any; quartz none at all; and mica,
which does not contain more than 8 or 10 *per
cent.*, occurs only in very small quantity in the
granite of Lusatia,—to which I here more particularly allude.

All these considerations prove beyond controversy, that basalt cannot originate from any
known rocks, altered and melted by volcanic
agents; in a word, that it is not, and never has
been lava.

It may here be objected to me, that a great
number of the lavas of Vesuvius, and especially of Etna, resemble basalt; that chemical
analysis has shewn, that both contain very
nearly the same constituent parts: and, it may
be inquired of me, Whence, then, have these
lavas proceeded? what rock has produced
them? I answer, Basalt itself. The rock of

Sicily, of a part of the south of Italy, of the Isle of Bourbon (Buonaparté), is basaltic: the volcanoes in these countries are, at least in part, situated in basalt; and it is that substance melted, which has furnished the material of their lavas. The comparative analyses of lavas and of rocks belonging to the basaltic family, (basalt, greenstone, and wacke,) published by Dr Kennedy, (§ 45), afford to this opinion all the authority which chemical analysis can bestow *. I have already admitted that the experiments of Sir James Hall prove, that melted basalt, if cooled very gradually, resumes in some measure the stony aspect. But even

H 4

---

* It is but fair to mention, that Werner taught this doctrine in his course of Geognosy, as early as 1786–7, and that he preceded all other mineralogists, or chemists, in remarking, that the substance of the lavas of Vesuvius, and especially of Etna, probably consisted in a great measure of melted basalt, seeing that the seat of these volcanoes was, at least in part, situated in the newest floetz-trap formation. In speaking of Vesuvius in particular, he said, that it was likely that the roof or upper part of volcanic caverns in general consisted of these trap rocks, while the lower parts might be situated in primitive rocks.   A.

though basaltic rocks, melted and manufactured by subterranean fires, should yield lavas
somewhat resembling such rocks *; it certainly does not follow, that these basaltic rocks
have originally been products of fire.

*Proofs derived from the heterogeneous substances
contained in Basalt.*

LET us now turn our attention to the foreign
matters found in basalt, and examine whether
their existence in that rock, and the state in
which they occur, can be reconciled with the
notion of its volcanic origin

The extraneous substances are. 1. Crystals
and grains, which often give basalt a porphyritic structure.   2. The balls and geodes which
partly or wholly occupy the air cells or vesicles.
3. Some fragments of older rocks.   4. Petrifactions.   5. Water.

41. I shall begin by reminding the reader,
that the crystals and grains most commonly

---

* NOTE O.

found in basalt, consist of basaltic hornblende, olivine, augite, mica, and felspar; that they exactly fill up the small cells which contain them; that they are nearly equally or uniformly disseminated through the mass of rock, and are scarcely ever observed to occur in groups.

In the opinion of volcanists, these crystals must either have existed previously to the lava, and have fallen into the melted matter; or they must have been formed in the midst of that matter, while yet in a state of fusion.

On the first supposition, it appears to me, that the crystals which could remain in the midst of the melted matter, would be found to consist only of the most refractory substances, such only being capable of withstanding the influence of the volcanic fire, joined to the dissolving action of the fluid which enveloped them. If these crystals had not been entirely melted, they would at least have been sensibly altered, as has happened to the crystals of augite and leucite, found imbedded in some of the lavas of Vesuvius. How then have crystals of felspar, of basaltic hornblende, of mica, and other

substances, been able to resist the fiery trial?
According to Saussure*, felspar is somewhat
more easily fusible than basalt; basaltic hornblende and mica, somewhat more difficultly.
Now, in the interior of masses of basalt, beyond the reach of weathering, basaltic hornblende is found of the finest black colour, and
the brightest lustre; its fracture, at the same
time, nowise vitreous, but completely lamellar:
crystals of felspar exist there, in all their freshness, and with all their translucency: those of
mica occur in plates, of a perfectly hexagonal
shape, and very thin, yet even the edges do
not bear the slightest marks of fusion. Olivine, it is true, is more refractory than these
substances; but its colour, which is very delicate, and very liable to be altered even by the
influence of the weather, is found in all its
native freshness, in the very middle of basaltic
rocks.

------

* Some naturalists, and Dolomieu among others,
consider basaltic hornblende as more fusible than felspar.   A.

In basalt, all these crystals most exactly fill up the small cells which contain them, so that they adhere to the walls of the cells; but in lavas, the crystals are often loose in the cavities, so as to make a rattling noise when shaken; in cooling, they have undergone a degree of contraction, a thing that happens to almost all bodies that have been submitted to the action of fire.

If the crystals which often occur in such numbers in basalt, had floated in this basalt while in a state of fusion, they would certainly have united and formed groups; we should have found them in knots or clusters; but certainly not *uniformly* distributed throughout the whole mass, each separate from its neighbour, and without any regard to difference in specific gravity *. In a word, these crystals are disposed in basalt exactly in the same manner as crystals are in all the porphyries, which are admitted to have been produced in the humid way.

---

* Note P.

Further, if it be considered that the extraneous minerals found in basalt, are almost always of the same kind; that they are every where basaltic hornblende, olivine, &c. in the Auvergne, and in Saxony, in Hessia as well as in the Isle of Bourbon; it is impossible to resist the conclusion, that it is not by chance that they appear there, but that they are as it were *indigenous* to basalt. It would seem as if they were placed there to testify to us, that all basaltic rocks have had the same origin; that they have all been produced by a single operation in nature; or in other words, that they have been precipitated from one and the same solution. When a volcanic origin is ascribed to them, it must be assumed, that the furnaces of all the volcanoes which have produced them, had been situated in the same species of rock, (which is contrary to what we know to be the fact, § 36.), and that that rock contained numerous grains of basaltic hornblende, olivine, &c. But there is not in our globe, any other rock than basalt, that contains these materials: it must therefore have been basalt that furnished the matter of basalt!—a conclusion which

does not very satisfactorily explain its first origin.

It may perhaps be alleged, that these crystals
did not exist within the great volcanic furnace;
but that they were situated in the rock which
the lava covered, and were thus enveloped by
it. In this way, no doubt, it is possible to account for some particular facts; but it is impossible thus to explain why the crystals are distributed in a uniform manner through the whole
mass, or why they are every where crystals
of basaltic hornblende, olivine, and augite.

We shall now more minutely examine the
opinion, that these crystals have been produced
in the matter of basalt, considered as lava,
while that matter was still in fusion, and that
they were so produced in consequence of affinities existing between their constituent particles. It must at once occur, that while the lava
continued to flow, and previously, the formation of the crystals could not take place. That
operation could only go on when the lava had
become stagnant, by entering some hollow,
like a reservoir. But even fluid lava is so glu-
tinous, that I have great difficulty in believing,

that the chemical attraction between the minute particles of the crystal disseminated
through the mass, could at all have overcome
the resistance offered to their approach to each
other by this viscidity;—far less that they
could have united, so as to produce crystals of
the most perfect form, and of a considerable
size. From all that we know concerning lava
in a state of fusion, it appears that it is very
thick, and very viscid. Large streams of lava
sometimes, it is said, take whole years to travel
over a short space of ground. Besides, how is
it possible to conceive, that felspar should crys-
tallize in the midst of melted basalt, while it is
itself more fusible than basalt, at least according to the tables published by Saussure. As
long as the mass remained in fusion, the caloric
would hold the minute particles of felspar, dis-
joined and separate from each other, and they
would assume the globular form. In some
stony matters which had been melted, I have
observed the crystals of felspar which they
contained, converted into globules of a white
enamel.—After all, I am not perhaps sufficiently acquainted with the effects of fire, to affirm

that what appears to me impossible, is so in reality *. I refer the question to the chemists.

One of these, (KLAPROTH,) whose name is well known in the literary world, who has studied minerals with the greatest assiduity, has had many opportunities of observing the effects of fire upon them, and who besides has examined, at different times, a great number of basaltic mountains,—Klaproth, I say, has given a decided opinion concerning the formation of basalt. His testimony being here of great weight, I shall extract a passage from his *Contributions* †, a work which his high esteem for Vauquelin, has induced him to dedicate to that philosopher. " The scope and limits which I have prescribed to myself in this memoir, (he observes), do not allow of my fully entering into the question concerning the origin of basalt, although the solution of that problem is of great interest in geology. I shall content myself with mentioning, that my individual opinion is

---

* NOTE Q.

† Beiträge, &c. vol. iii.

the result of observations which I have myself
made on many basaltic mountains; and that
these observations have convinced me, that basalt is a stony mass, produced in the humid
way." In treating of Porphyry slate (Porphir-
Schiefer,) a rock which has much affinity with
basalt, he says, " The opinion which prevailed
some time ago, concerning the origin of porphyry-slate, and which has still some partisans,
ranks that substance, as well as basalt amygdaloidal rocks, and others belonging to the
trap formation, among the products of volcanoes, or, in other words, among lavas. It is
not here my intention to rehearse and discuss
all that has been written for and against that
opinion. I shall merely state, that the observations which I have made at different times
among the mountains of Bohemia, on the position and mineral relations of the rocks of basalt and porphyry-slate, have never led me,
more than other unprejudiced observers, to discover the least vestige of a crater, or any other
indication of volcanic influence." I may here
remark, that the country in which Klaproth
made these observations, is one where, of all

others, imagination might trace at every step, supposed marks of the action of fire. As happens in some parts of Auvergne and of Italy, the soil is blackish, and contains interspersed crystals of basaltic hornblende, augite, melanite, and balls of basalt; while the mountains consist of groups of conical eminences, having a smoked appearance.

42. There may often be observed in masses of basalt, a number of cavities or vesicles, generally of a round shape, some empty, and some occupied wholly or in part with balls or geodes of green-earth, of steatite, calcspar, zeolite, calcedony, quartz, and other substances. These *balls* differ from the *grains* mentioned in the preceding paragraph, in being evidently posterior in formation to the basalt in which they are found *. The lavas of Vesuvius or Etna, do not contain any similar balls : the cavities in these lavas bear no marks even of the commencement of their formation. Amygdaloidal

I

* See Note A.

masses, it is true, are frequently talked of as
volcanic products ; but those who hold this doctrine, begin by taking for granted that basalt,
wacke, and other rocks which are only modifications of the same substance, have had a
volcanic origin. I shall here mention two instances of amygdaloidal rocks, which it is absolutely impossible to regard as congealed
streams of lava. In Derbyshire, in England,
there is found a substance called *toadstone*,
which is nearly related to basalt : it contains a
great number of nodules of calcspar : it occurs
in beds, which alternate with beds of *limestone*,
in a district traversed by metallic veins. The
aqueous origin of such beds is evident. In the
county of Glatz, I examined a hill *, the summit of which consisted of a sort of wacke, of
a reddish-brown colour : it was full of nodules
of calcedony, generally coated with a layer of
green-earth. Yet, that that wacke must have
been produced in the humid way, was incon-

---

* The author alludes to *Finkenhübel*. For a more
particular description of the hill and of the wacke which
contains the turbinites, see Von Buch's description of
the Environs of Landeck, p. 96. of Dr Anderson's
translation.   T.

testibly proved by its containing imbedded *tur-binites*. To put the fact of the existence of these petrifactions beyond a doubt, M. Von Buch has deposited a specimen of the rock in the cabinet of natural history at Berlin *.

43. Basalt has been observed to contain but few fragments of rocks which existed anterior to its formation. For my own part, I have seen only some fragments of sandstone. But other naturalists have discovered several substances, chiefly calcareous. Werner, in a memoir printed in the *Journal des Mines de Freiberg*, informs us, that in the environs of Carlsbad in Bohemia, the basalt contains so great a quantity of calcareous matter, that the people make lime from it. Saussure, in his Travels in the Alps, § 1611, mentions his having seen basalt containing angular frag-ments of a grey compact limestone, and these fragments in nowise altered at the point of contact. Is it possible to suppose, that a torrent of melted stones should envelop frag-

---

* *Journal de Physique*, vol. 47.

ments of carbonate of lime, without producing the slightest alteration on their edges, or
the least disengagement of carbonic acid
gas? Yet the edges are unaltered; and any
disengagement of gas must have caused some
blisters or vesicles in the adjacent fluid
mass; but none such are to be seen. Since
this carbonate of lime exists in vast quantity, it
evidently has not merely been enveloped by a
stream of lava accidentally coming in contact
with it; and it is needless to enlarge on the
impossibility of its having been *formed* in
the midst of stony matter in a state of fusion.
In the instance mentioned by Werner, the calcareous matter does not consist of fragments,
but of portions which have been formed at the
same time with the basalt.

Among the *ejected stones* of Vesuvius, it may
be remarked, many fragments of limestone occur; but they are never found (at least according
to what I have read) in its lavas, which always
appear of a homogeneous texture. These fragments are probably the debris of the walls of the
volcanic caverns: many such must inevitably
have fallen into the great volcanic furnaces, and
have been melted;—affording satisfactory proof,

that the heat of volcanoes is capable of altering, and even melting limestone, and that that substance could not remain unaltered in the midst of lava.

44. I have already (§ 42.) taken notice of the turbinites found by Von Buch in a rock of the trap-formation, in the county of Glatz. Petrifactions, however, are of rare occurrence in the basaltic rocks of Germany; and as I do not wish to be much indebted to the observations of others, I shall only cite two or three examples of their occurrence elsewhere. Dr Blagden, secretary of the Royal Society of London, and Mr Chenevix, inform me that they have seen impressions of shells in the hardest and most compact basalt of the Giant's Causeway in Ireland [*]. In the Vicentine, behind the hill of Carlsberg, in the Veronese, chamites occur in basalt. Von Berolding describes a species of cornu ammonis, found in basalt at Forez, the shells still retaining their pearly lustre. He

---

[*] It may be proper to notice, that the rock at Portrush, near the Giant's Causeway, here alluded to, is not basalt, but a variety of Lydian-stone, which sometimes approaches to slate-clay.   T.

likewise speaks of two pieces of basalt, from
the neighbourhood of the lake of Constance,
which contain gryphites and ammonites.
Brugnatelli mentions basalt from the valley of
Ronca, containing sea-shells. At Joachimsthal,
in Bohemia, there is a great mass of wacke,
in which are found entire trees, in a half petrified state, but still retaining their bark, their
branches, and even their leaves *.—It would
surely be unnecessary to enter into any argument, to shew, that the mineral substances in
which these petrifactions are found, cannot
have been the products of fire; yet these substances are either basalt, or modifications of that
rock, which must have had the same origin.

45. The circumstance of water being sometimes found inclosed in the vesicular cavities of
basalt, might be urged as a strong objection
against its igneous formation. But as it is impossible for me to demonstrate, that the water
may not have penetrated by infiltration, and as
I have never had an opportunity of seeing the

---

* Von Buch's *Beschreibung von Landec.*

fact with my own eyes, I shall not insist on it. I must, however, particularly remark, that ba salt contains a very considerable portion of water of *composition*, while lavas do not con tain any, as may be seen from the following analyses made by the late Dr Kennedy * :

	Basalt of Staffa.	Greenstone or Whinstone of Salisbury Craig.	Wacke of Calton-hill, Edinburgh.	Lavas of Ca tania, Etna.	Lava of St Venere, Etna.
Silica, ...	46	46	50	51	51
Alumina, ...	16	19	$18\frac{1}{2}$	19	17
Lime, ...	9	8	3	$9\frac{1}{2}$	10
Oxide of iron, ...	16	17	17	$14\frac{1}{2}$	$14\frac{1}{2}$
Water and volatile matters, ...	5	4	5	0	0
Soda, (about *)*	4	$3\frac{1}{2}$	4	4	4
Muriatic acid, (about)	1	1	1	1	1
Loss, ...	3	$1\frac{1}{2}$	$1\frac{1}{2}$	1	$2\frac{1}{2}$

Proofs derived from the form of basaltic rocks.

46. Basaltic rocks are generally found ar ranged in regular beds, which often at the same time exhibit the prismatic division. (Note A). I have already remarked, (§ 35.)

---

* *Annales de Chimie*, ventose an 10.

that streams of melted matter would never take the form of thin beds, particularly of a great extent, and presenting alternate risings and sinkings, such as occur in beds of basalt. I have hitherto alluded only to the great bed of basalt which covers the mountains of Saxony: but even this becomes comparatively insignificant, since Dolomieu describes beds of basalt occurring several hundred yards below calcareous mountains,—and no fewer than twenty beds of basalt alternating regularly with as many of limestone; and since Professor Jameson informs us, that in the Western Islands of Scotland, great beds of sandstone occur, having interposed between them beds of basalt, sometimes only a few inches in thickness.   One of two things is certain : these beds must either be *waved*, (which is the ordinary case with all beds), and if so, they could not, we have seen, have arisen from streams of lava; or they must be *flat*,—and that is certainly not the form which a stream of matter in fusion would assume.

As to the prismatic division which basalt affects, I believe that if it is found in the true lavas of Vesuvius or of Etna, it is only in a

very imperfect way. I know that some authors have spoken of the columnar shape
as a distinguishing mark of volcanic productions; but they begin by taking for granted,
that basalt is a volcanic production. I have
never heard it alleged, that any lavas which
have flowed in the memory of man, have assumed the form of perfectly regular prisms,
sonorous like iron, such as the columns of Stolpen, of the Heulenberg, (§ 21, 20), and others.

*Proofs from the position of basalt with respect
to other rocks.*

47. WE must examine the relations of basalt, with rocks of the same *formation,* and
with those also of a different *formation.* I trust
I have satisfactorily shewn, (§ 29, 30, 37, &
38), that the relations of basalt with rocks of
the former description, do not admit of the
supposition of a volcanic origin. As to rocks
of a different formation, whether the basalt
covers them, which is almost always the case,
or whether it be itself covered, we can only
draw conclusions from the relative position,

and from any peculiarities observable at the points of junction.

Wherever I have been able distinctly to see the place where basalt rests immediately on granite, porphyry, limestone, or clay, I do declare that I have never perceived the smallest alteration on these rocks,—nothing that would indicate that they had been covered by a stream of matter in fusion. In many places, basalt lies immediately over beds of coal; and yet, at the surface of contact, none of the effects of fire are discernible, although that bituminous substance must necessarily be very sensible to the action of that element. The Meisner hill, in Hessia, (Note C), presents a remarkable example of this singular kind of superposition. It is true, indeed, that in this hill, the bituminous bed is generally separated from the basalt, by a thin bed of clay : but having explored above 9000 feet of the subterranean excavations in that mountain, I have had an opportunity of seeing the basalt and coal in immediate contact in several places. At the points of junction, there occurs sometimes an earthy matter, impregnated with a good deal of bitumen, and sometimes a coal of a very resinous ap-

pearance, divided into little bars*, and entirely of the same nature as that which isfound
under the clay.   Now, if the basalt which
forms the roof of this bed of coal, and which
is more than 300 feet thick, had ever been in
the state of a vast stream of melted stones ; is
it possible to suppose, that in thus spreading
itself over a bed of bituminous substances, it
would not in any manner have altered them ?
I do not go the length of saying, that such a
fiery stream must have inflamed and consumed
the coal, because it may be objected to me,
that all communication with the atmospheric
air was instantly cut off; but I affirm, that it
must at least have produced some sort of alteration.   It appears to me impossible to conceive, that a bed of very bituminous coal
should be no otherwise affected by a red-hot
stream of melted stones, than by a simple sediment of clay !

In the Meisner, the basalt lies between a bed
of coal, and a rock with a granular structure.—
Is this the kind of repository in which lava
should be expected?

---

* Stängenkohle, *Werner.*   Columnar Coal, *Jameson.*

In that hill, the basalt is merely found resting on the coal, which is chiefly of the species called Brown-coal, (Braunkohle). But in other places the basalt is found alternating with the coal in beds. Mr Williams informs us, that at Borrowstounness in Linlithgowshire, Scotland, there are thick beds of basalt situated between beds of coal which are worked to a great extent; and that in the Bathgate hills, in the same county, these two mineral substances may be seen alternating in beds. In Bohemia, also, according to Dr Reuss, beds of coal are worked, which are situated between beds of basalt. In the Feroe islands, examples of a similar arrangement have been observed. Mr Jameson, who is both a skilful mineralogist and a correct observer, says, " We noticed in the north-east part of the island of Mull, a bed of coal, a foot in thickness, the *roof* of which consisted of basalt, irregularly divided into columns. The same substance formed also the *floor* of the coal *."

---

* Outline of the Mineralogy of the Scottish Islands, Edinburgh, 1800.

For some observations on the occurrence of alter-

I can scarcely think that any one will still maintain the proposition, that those basaltic rocks which are thus arranged in alternate beds with bituminous and combustible substances proceeding from the decomposition of organic matters, have originally consisted of burning streams of melted stones.—For my part, I cannot even conceive the existence of such a phenomenon : to suppose it, seems not merely to make Nature deviate from her ordinary course, but to make her transgress the bounds of possibility.

The Meisner hill is the only place where I have myself seen basalt covered by another mineral substance : but various naturalists describe similar facts. They have found basalt covered by, and even alternating with, shelly limestone, sandstone, and slate-clay. Professor Jameson, whom I have just quoted, details a fact of that kind, which he saw in the most distinct manner in the island of Eigg, on the west coast of Scotland *.

---

nating beds of basalt, or of greenstone, in Linlithgow-shire, Mull, and in the Feroe Islands, see Note R. T.

* Note S.

Dolomieu observed, in Auvergne, beds of basalt alternating with limestone containing impressions of shells. In the Vicentine and the Tyrol, he found twenty thick beds of basalt, interposed between beds of limestone*. The volcano which vomited forth these lavas, must no doubt have been a submarine one, since the basalt alternates with shelly limestone. But to me it appears, that matter in a state of fusion, would scarcely spread itself quietly and uniformly under the water, so as to form a wide horizontal bed, but would much more probably swell up and become heaped around the crater. If it be still objected to me here, that it is impossible to compare the great operations of nature, with those of our laboratories and furnaces, and that the effects of a river of lava cannot be judged of, from merely seeing a streamlet of scoriæ issuing from a furnace; Dolomieu himself shall furnish the answer. " I cannot possibly believe," he says, " that streams of lava, many leagues in extent, can have flowed under the waters of the sea.—At

* Note T.

Etna and Vesuvius, the lavas become heaped to a great height, instead of spreading to a distance. Indeed it is impossible that they could continue long to flow under water, as the water would necessarily tend to coagulate them, and deprive them of their progressive motion." On the volcanic hypothesis, also, the regular alternation of twenty eruptions of lava, with twenty depositions of a calcareous sediment, must appear very surprising; and it certainly will not be easy to explain how streams of red-hot melted matter have happened to spread themselves over carbonate of lime, without producing any alteration whatever upon it. Some naturalists, in order to explain this alternation of beds of basalt and of limestone, have supposed that the volcano was not submarine, but that after each of the twenty eruptions, the sea rose and covered the lava; and that after depositing its calcareous sediment, the water again retired. This playing of cards between Vulcan and Neptune,—this alternation between the eruptions of a volcano and the disappearance of a whole sea,—was repeated twenty times!— But the period has now arrived, when geolo-

gists must submit to be stripped of the right
which they have arrogated to themselves, of
commanding Nature, and ascribing to her, at
their pleasure, the most extraordinary and fan-
tastical movements; making seas abandon their
limits, transporting their waters whither they
please, and ordering them to come and go, as
suits their conveniency;—and all this to explain
the merest hypotheses.

When we see beds of different substances
alternating with perfect regularity, and, by
their assemblage, constituting mountains which
do not exhibit the marks of having undergone
any extraordinary convulsion or alteration, it
would surely be a violent supposition to insist
that some of these beds have had an origin of
quite an opposite nature from others; that while
one set of them evidently consists of sediments,
or aqueous products; another set, interposed between the former, consists of igneous products,
or congealed streams of lava.

48. The conical shape of basaltic mountains,
which has contributed so much to their being
taken for extinct volcanoes, is of itself suffi-

cient to prove the falsity of the preceding hypothesis. I have satisfactorily shewn, (§ 34.)
that this shape is not that of volcanic mountains, but in some measure opposed to it; and
that even when we do not, by means of galleries
pushed under the basaltic summits, or by deep
wells sunk through the basalt, and terminating
in gneiss *, know with physical certainty, that
the basalt is only superimposed,—the very aspect of the mountains on which it occurs, is of
itself a convincing proof of the fact. I shall
relate still another example. Near to Kirchen,
about fifty miles to the north-west of Coblentz,
in the midst of a mountainous country, where
the prevalent rock is argillaceous shistus, (grey-
wacke slate), a large mountain rises to some

K

---

* Not far from Friedberg in Silesia, there is a basaltic hill called Greiffenberg, or Greiffenstein. On
this hill, a well has been dug more than a hundred feet
deep. For about one-half of this depth, the rock continued to be basalt: it then suddenly became gneiss.—
When in Silesia, I myself went to Gieren, where the
warden of the Friedberg mines resides, and had the
fact confirmed to me.   A.

height above all the rest, of an oblong shape, and nearly twelve miles in circumference. In the middle of its highest ridge, which is of the *dos d'âne* form, a basaltic eminence occurs, of a small size considering the magnitude of the body of the mountain : it may be near seventy feet high, and perhaps more than a thousand in circumference. The rock-beds of the mountain are inclined to the horizon at angles of 70° or 80°. I traced and examined the stratification up to the base of the basaltic eminence; but I could not perceive any derangement. I extended my examination all around the eminence, and on both sides ; and every where, the beds had the same direction, and the same inclination : from which I conclude, that they pass entirely through below the basaltic summit, without undergoing any derangement. Certainly this mass of basalt must formerly have been of much greater extent : the destructive influence of time is continually diminishing its size, and laying bare more and more of the nucleus of the mountain, which the basalt previously concealed. Perhaps, some ages hence, no vestige of basalt may here remain:

the mountain will then appear to be wholly composed of grey-wacke slate; and the naturalists
of those times will learn with astonishment
from history, that, in the nineteenth century, it
was looked upon as a volcano.

The reader may probably be surprised, that I
have not yet taken any notice of the supposed
vestiges of craters, more or less distinct, which
some observers persuade themselves they are
able to discover on all basaltic mountains. The
truth is, that in those which I have had an opportunity of examining, I perceived nothing of
this sort. The only hollows I could find, were
either excavations evidently made by the hand
of man, or which owed their origin to some ordinary cause, easily detected. Even a pre-
judiced observer of the most lively fancy, could
scarce convert such hollows into volcanic gulfs
or abysses.

49. If we examine the topographical position of the basaltic mountains of Germany
considered as a whole, we may perhaps be able
thence to draw some inferences concerning the
origin of basalt in general. I shall here give a

K 2

rapid sketch of the situation of those basaltic
mountains which I myself have seen.   Germany
is traversed, and in some measure divided by a
grand chain of mountains, (the *highest mountain-chain* of the country), which takes its rise
towards the Crapak or Carpathian Mountains.
After sweeping along Silesia, Lusatia, Bohemia,
Saxony, it joins the mountains of Fulda, and of
Hessia, and touches the Rhine between Mayence
and Cologne.   This chain forms the *ridge* of the
extensive country situated between the bed of
the Danube and the northern seas ; (or it is the
highest mountain-ridge of Germany).   The summits of a great number of the mountains which
compose this elevated saddle, are of basalt ; and
that mineral is scarcely found elsewhere in the
rest of Germany.   So that it may be said, that
in this part of Europe, the basaltic mountains
form as it were a train or series placed exactly
on the ridge of the country, comprised between
the Danube and the northern seas.

Nearly the whole of this highest mountain-
ridge is composed of granite, of gneiss, and of
mica-slate ; in short, the rocks are primitive,
(the limestone of Hessia, and a few other rocks

forming exceptions). If I have not already
succeeded in shewing that these basaltic mountains have never been volcanoes, I may urge
that it must appear a very singular circumtances,
to find a series of such craters in the most elevated part of the German empire, and in the
midst of primitive rocks.—If it be here objected
to me, as it possibly may, that the chain of the
Cordilleras presents a similar phenomenon, I
can only answer, that I am not possessed of
sufficient knowledge of the topography and
geology of that chain, to enable me to form a
correct opinion on the subject.

50. Such are the principal reasons which
have convinced me that the basaltic rocks *which
I have hitherto had an opportunity of seeing*, have
not resulted from subterranean fires.—My purpose here, has not been to treat fully the question concerning the origin of basalt in general : I have not advanced any thing not founded
on my personal observations ; and I have only
deduced such conclusions from them as appeared to be inevitable. Perhaps at some future day,—after I myself may have viewed

volcanoes and their products, and examined the basaltes and extinguished volcanoes of Auvergne and the Vivarais,—perhaps I may then be better fitted to consider that question in all its extent,—to form a just estimate of what has been written on the subject, and to bring forward more interesting facts and examples. One of my fondest wishes is to undertake such a journey, after completing the preparatory examinations which now engage me*. There can scarcely be more interesting problems to French naturalists, than to determine if so great a portion of their country has really at one time been the theatre of the magnificent and awful scenes presented by volcanoes; if the soil

---

* Mr Daubuisson accomplished his visit to Auvergne in the following year ; and read an account of his journey and remarks to the National Institute. (*Journal de Physique*, t. 58). He now agrees with Dolomieu and other naturalists in regarding the *puys* or conical eminences of Auvergne, as extinct volcanoes, and the basalt of that country, as of volcanic origin, or basaltic lava ; but he has in nowise altered his opinion concerning the aqueous origin of the basalt of Saxony. Some account of his observations will be found in NOTE *u*.

T.

which is now cultivated has, at some remote period, been an ocean of fiery matter.—We are now aware of the kind of facts which must be resorted to in solving such a problem.

PART V.

CONCLUSION CONCERNING BASALT IN GENERAL.

§ 51. LET us take a bird's-eye glance of the extent of the basaltic mass in general, that is, of the mountains of basalt which occur in different parts of the globe. We shall begin with those of Saxony. To the east, they unite with those of Lusatia; from thence they join with those of Lower Silesia, of the county of Glatz, and of Upper Silesia; and they continue as far as the Carpathian Mountains. All of these mountains form only, as it were, a single great chain. Proceeding still farther to the eastward, basalt occurs in Armenia, in Persia, and to the

very extremities of Asia.   Hindostan, the Isles
of the South Sea, the *ci-devant* Isle of Bourbon,
contain basalt in great quantity.   To the north-
ward of Saxony, we find no more in Germany,
the country being low and flat : but, on the
other side of the Baltic Sea, it re-appears in
Sweden, and in Norway; Scotland, Ireland,
and the Orcades *, are full of it.   To the west
of Saxony, we find it again in Franconia; but
more particularly, in directing our views a little
towards the north, in the province of Fulda, in
Hessia, and the country bordering on the Rhine.
Beyond that river, we discover some traces of it
in Burgundy; and in Auvergne, and among the
Cevernes, it occurs in vast quantity.   Proceeding
still farther west, we find it again in Spain and
Portugal.   To the south of Saxony, all the

---

* M. Daubuisson perhaps employs the term " les
Orcades," for the Islands of Scotland in general, or he
may have inadvertently written it, in place of *Hebrides*.
In the *Orkneys* properly so called, basalt, if it occurs
at all, is found very sparingly.   It is likewise uncommon in Shetland.   But in the Hebrides, it is very
abundant.   T.

mountains of the northern part of Bohemia, are covered with that substance. A little occurs in Stiria, and in Carinthia. I do not know if any be found in the Alps : but beyond that, in Italy, basalt appears in the Vicentine, in some parts of the Appennines, and, according to my information, it seems to constitute the rock of a part of the south of Italy. It makes its appearance among the Islands of the Archipelago, and extends even into Africa ; the antique basaltes having been brought from Ethiopia.

When I attentively consider these different ramifications of the great basaltic mass, thus to be found on our globe * ; when I observe how uniformly they incline to contiguity with each other ; when I find, (from ocular inspection of mountains and of cabinet speci-

---

* The view which I have here given of the extent of the basaltic mass, must be regarded merely as a loose sketch, very far from complete. The truth is, the greatest part of the surface of the earth has not yet been surveyed by mineralogists; and my acquaintance with mineralogical geography is but limited. **A.**

mens, as well as from what I have read), that
basalt is really every where the same substance,
—that it has the same black colour,—the same
earthy fracture, very small granular or compact,—the same specific gravity of 3.000,—the
same hardness,—the same tendency to prismatic division,—that it occurs every where over
the surface of the globe, in scattered patches,
forming generally the summits of some lofty
and conical hills,—that it contains very commonly the same sort of basaltic hornblende
and olivine,—that it is almost always accompanied by greenstone, by wacke, or by porphyry-slate,—and that, by analysis, it affords
very nearly the same results; when I consider
these facts, I cannot help regarding the basalts
of different countries, as parts of one whole,—
as partaking of one common origin,—as mere
branches of the same stock.   I do not entertain
any doubt concerning the aqueous formation of
those which I myself have had an opportunity of examining; and, led by analogy, I am
inclined to believe that others, possessing the
same identical characters, have not been produced in a different or opposite manner.

I must, however, remind the reader, that this last conclusion can be regarded only as a *conjecture hazarded*, as I announced in the beginning of the memoir, (§ 3). It has been suggested to me only, by reading descriptions, and examining cabinet specimens. On such data, it is impossible to form more than conjectures. In natural history, it ought to be a rule, that thorough conviction should result only from personal observation.

Perhaps the same defect may be ascribed to my reasoning, which has been objected to the argument of certain geologists, who, very willing to have it concluded that all basalt is of volcanic origin, reason thus: " We are sure that the black masses found on Vesuvius, Etna, &c., are products of fire, since they form a part of the streams of lava ;—we find in other places, black masses resembling the former ;—these must therefore have had the same origin." To this I would answer, that although the resemblance may be striking in some respects, it is far from being so in all, especially in the circumstances attending its position or *repository*.

Having brought forward a collection of well-authenticated facts, and stated such conclusions as seemed naturally to result from them, I have permitted myself to indulge only in a single conjecture. If it be allowable to compare small things with great, I may observe, that M. de LAPLACE, concludes his " Mechanique celeste,"—a work which displays the most accurate and most sublime conceptions of the human mind,—with a *conjecture* concerning the origin of the planetary system; which, he declares, he " submits with that diffidence which ought to accompany every opinion which is not the immediate result of observation or calculation." If a philosopher, who is so intimately acquainted with the course of nature, and whose speculations all bear the stamp of the highest probability, proposes with diffidence a conjecture which corresponds entirely with the known and inflexible laws of the heavenly bodies; I am sensible how little confidence is due to that which I have stated concerning the origin of basalt in general.

52. I may perhaps be required to say, whether there are not basaltes which have actually been the product of volcanic eruptions; and whether I regard as fabulous the accounts which so many distinguished naturalists have given of extinguished volcanoes? I answer, that since some new volcanoes have broken out within the memory of man, it is highly probable, that others may have become extinct. I do indeed believe, that such extinct volcanoes exist, although I have never seen any of them. Further, it seems very likely, that they should occur in a basaltic district, since the greater part of active volcanoes are situated in such rocks. Without entering into any discussion concerning the cause of the phenomena presented by volcanoes, or the nature of the combustible substance which supports the fire, I can easily conceive, from what I know, the possibility of the existence of *basaltic lavas.*

I have seen beds of coal on fire below beds of basalt. The heat arising from such combustion, is sufficient to melt the basalt. Some particular circumstances, as the introduction of a great quantity of water into the midst of the fused mass, might here occasion an eruption; streams of lava would be produced; and these, on cooling, might in part assume the stony aspect of the basalt from which they were produced. Even without supposing a volcanic eruption, it might very well happen that the basalt melted by the combustion of the coal, finding some passage for escape, might spread itself over a lower piece of ground. It would in this case form a stream of lava, one portion of which might, when cooled, have a basaltic appearance, while another might consist of true scoriæ, with roasted and calcined stones resembling pumice.—There may thus be found in basaltic districts, true lavas or streams of melted substances, among which basalt may occur, which, after fusion, has resumed, to a certain degree, its previous appearance: it is still true basalt, since it has not changed its nature, or become a different substance;— in the same way as it may be said, that melted gold is not the less gold after it has cooled.

But let us examine the difference between those basalts which remain in their original position, and these basaltic lavas, or *lava-basalts*, as they may be called. These last can never contain crystals of felspar possessed of its usual lustre and translucency; they can never be found constituting vast beds, which cover entire countries, and extend indifferently over hills and valleys; nor spreading over beds of coal, without producing the least alteration; nor alternating twenty times with beds of limestone! I make my appeal to every student of physics and natural history, whether, when a trifling patch of basalt is found upon the summit of a great mountain of granite or mica-slate, in which no excavation or laceration is to be seen,—whether it is supposable that this basalt could have issued from the bowels of that mountain?—Again, when a single thin bed of basalt covers a whole chain of mountains, rising and falling according to the nature of the surface; when it forms a continuous mass, without affording the least vestige of those scoriæ, rapilli, and cinders, never awanting in the vicinity of volcanoes; whether such a bed could possibly be the product of a volcanic eruption? —And once more, when a vast horizontal platform of basalt is found, perhaps three leagues

in length, a league in breadth, and above three
hundred feet thick, having the upper part of a
crystalline texture, and the base resting immediately on a bed of bituminous and inflammable substances, which have not thereby undergone the slightest alteration; whether it is
possible to conceive that such a basaltic platform could ever have been a sheet of melted
stones?—I appeal to chemists, whether a rock
which presents at every step, disseminated in a
uniform manner through its substance, crystals
of felspar possessed of their usual lustre and
translucency, can possibly have been a product
of fire,—an agent which so readily destroys or
alters that mineral?

I think it needless to extend farther the
enumeration of such arguments; any one of
which, by itself, appears to me decisive of the
question.

53. If the hypothesis, which supposes the basaltic rocks of Saxony and some others to be
volcanic products, is so completely at variance
with the results to which correct observation
leads, and with the principles of sound physics ;

it may naturally be enquired how the former opinion has gained such currency. I would ascribe it in a great measure to a love for the marvellous. It is well known how readily any thing that pleases by striking the imagination, is received, and with what difficulty the simple unaffected truth can procure a hearing. It delights the imagination to suppose the existence, in former ages, of burning mountains, and vast streams of liquid fire,—nature, as it were, in a conflagration,—at the very spot, perhaps, now occupied by our peaceful habitations. From our childhood, we are charmed with the pompous description of the great and terrible phenomena of volcanoes *. The imagination pictures to itself the appearance of the fiery gulfs; and

---

* How forcible is the picture which Virgil has given of the phenomena of Etna:

.... horrificis juxta tonat Æthna ruinis,
Interdumque atram prorumpit ad æthera nubem,
Turbine fumantum piceo, et candente favillâ;
Attollitque globos flammarum, et sidera lambit;
Interdum scopulos, avulsaque viscera montis
Erigit eructans, liquefactaque saxa sub auras,
Cum gemitu glomerat, fundoque exæstuat imo.
                                        ÆNEID, lib. 3,

L

is wonderfully pleased with the employment
of tracing the contrast between the present
and former state of supposed volcanic mountains. The black and fuliginous aspect of basaltic masses, and the isolated and striking
shape of the mountains which they constitute,
greatly aid the illusion.

The resemblance between basalt and certain
products of volcanoes, either lavas or ejected
substances, naturally excites the notion of its
volcanic origin. It seems therefore nowise extraordinary that such an opinion should have
been advanced at different periods; nor was
there any want of specious arguments in support
of it. Some naturalists having afterwards discovered the products of subterranean fires, and
likewise some extinguished volcanoes, in the
midst of basaltic mountains, were induced by a
general similarity of aspect, and by taking a part
for the whole, to ascribe to the action of fire the
formation of all black stony masses; while the
truth is, that only a small portion had been re-
moulded by that element. The line of distinction, according to the report of observers, is
not always indeed easily assigned. But, I

repeat it, it is the love of the marvellous which has given currency to this opinion among the less informed class of society ; (for, it will be observed, that I have not accused philosophers or physical inquirers, of being influenced by such a principle, but have only said, that it was *very natural* for them to be misled by first observations). Let the reader reflect, that in the Vivarais, the flame of peeled hemp-stalks, used as lights during the summer evenings, by the villagers engaged in winding silk from the cocoons, has by travellers been magnified into volcanic fire * !

54. It appears, says the celebrated chemist of Berlin †, that naturalists are recovering by

---

* Chaptal, in his Elements of Chemistry, (article *volcanoes*), mentions this circumstance, and gives a lively account of the proceedings of a party of philosophers who, equipped with torches and speaking-trumpets, determined to examine the phenomenon more closely. On arriving at the spot, they were saluted, by the affrighted country people, with a shower of stones,— which Chaptal quaintly hints they might possibly mistake for a volcanic eruption.

† Klaproth, *Journal des Mines*, No. 74,

degrees from the volcanic illusion. It is about
fifty years since a French naturalist revived
the opinion concerning the volcanic origin of
basalt, and he lived to see almost all Europe
adopt his sentiments. BERGMAN, the first of
the chemists who employed himself with dili-
gence and success in examining mineral substances, and who, to an intimate acquaintance
with the effects of heat, joined an extensive
knowledge of mineralogy ; could not bring himself to consider basalt as a product of volcanic
eruptions. The Swedes adopted his view of the
question. It is scarce forty years since every
body in Germany considered basaltic mountains
as ancient volcanoes. WERNER lifted the Neptunian standard ; and now, among all the German mineralogists of any reputation, I know
but one (VOIGT) who still maintains the old
doctrine. We have already seen (§ 41.) in how
decisive a manner KLAPROTH has pronounced on
the subject : he, of all the German chemists,
has had most opportunities of observing the
effects of fire on mineral substances, and he has
besides studied the history of basaltic mountains with that correctness for which he is remarkable. In Ireland, Mr KIRWAN was a sup-

porter of the volcanic doctrine; but the numerous chemical experiments which he made on minerals, and other considerations, led him to a change. Dr MITCHELL*, one of the very best mineralogists, and Mr JAMESON, the author of the Mineralogical Travels in Scotland, and the greater part of British naturalists, consider basalt as having been produced in the humid way.

The geologist, who of all others possessed the greatest experience,—SAUSSURE, the illustrious mineralogist of the Alps,—found it necessary, in the latter part of his life, greatly to limit his notions as to basaltes being of volcanic origin. In speaking of the extinguished volcanoes of Brisgaw, he says †, " I acknowledge, that before studying the writings of Werner, I felt no hesitation ; but that philosopher has

---

* The late GEORGE MITCHELL, M. D. of Belfast in Ireland. He studied under Werner at Freyberg, between the years 1797 and 1802. He died at London in 1803, aged little more than forty years. For a character of Dr Mitchell as a mineralogist, see Introduction to Professor Jameson's Mineralogy, first edition.

† *Journal de Physique,* an 2, p. 356. T.

taught me to doubt." Dolomieu, who was at
the head of the Vulcanists, but in whom the
love of truth was paramount to the spirit of party, admitted that some basaltes had been produced in the humid way. He observes[*], " I
have circumscribed the volcanic empire more
than any other mineralogist, French, English,
or Italian, having withdrawn from its dominion
many mineral substances formerly placed under
it. I hold that the basalts of Saxony, of
Scotland, and of Sweden, may claim a Neptunian origin." When treating of the basalt of
Ethiopia, he adds, " I may affirm with certainty, that it is not of volcanic production."

From the detail just given, it appears, that
all the celebrated foreign mineralogists, and
the Swedish, German and English chemists,
who have paid most attention to mineralogy,
and have had opportunities of inspecting basaltic mountains, have not thought themselves
warranted to believe that basalt is a product
of fire, or has once been in the state of lava.

It may seem presumptuous to anticipate

---

[*] *Journal de Physique*, tom. 37.

the result of the discussion which is now agitated concerning the aqueous or igneous origin of basalt, or to predict what may probably, in a short time, be the opinion of all philosophers on that subject. But if it be allowable to draw conclusions from analogy, I would argue thus. It is only forty years since all the mineralogists of Europe considered basalt as of undoubted volcanic origin : at this day, the greater number of these, have abandoned that opinion : and it will not be disputed, that when, in a question connected with physics or natural history, a number of philosophers of the first character, who have long supported a particular opinion, see cause to change their views, they will seldom be found to be in the wrong. I may further remark, that when the Newtonian school, armed with all the force afforded by mathematical precision and certainty, attacked the system then generally adopted concerning the formation of the world and the movements of the heavenly bodies, the Cartesians began to modify their vortices in different ways ; and that when the French chemists, after carefully observing what took place in the phenomena of

chemistry, and excluding from their explana·
tions what was only hypothetical, undertook to
give a new form to that science, the Stahlians
were as ready to modify their phlogiston,
bestowing on it new attributes and various
forms. On the subject more immediately
before us, we see naturalists and chemists in-
troducing into their observations, a scrupulous
accuracy hitherto unknown in geology; they
are calling in question the igneous origin of
basalt; and the Vulcanists are modifying their
opinions, and talking of agents with which we
are unacquainted, and which are new in their
nature. When the analogy between the circumstances attending this discussion, and those
which accompanied the decline of the Cartesian
and Stahlian doctrines, is considered, it seems
natural to conclude, that the result may probably be the same.

55. I shall refer to the notes (Note X.) the
examination of the hypothesis concerning *vol-
canisation*, proposed by DOLOMIEU, the naturalist who, of all others, has distinguished himself by studying volcanoes.

56. It may possibly be insinuated, that I have been prepossessed or influenced in my observations of nature, as well as in the management of this discussion, by a partiality for the doctrines of the school where I imbibed the rudiments of geological knowledge. But I must remark, that if I had been liable to be influenced by such prepossessions, the national prejudice would greatly have overbalanced that of the school. I knew that the greater part of French mineralogists considered basalt as a volcanic production : my predilection for those philosophers, and the peculiar appearance of the first basaltic rocks which I had an opportunity of seeing, led me naturally to incline to their opinion. But when I had made a greater number of actual observations,—when I had examined with care the nature of basalt, its connection with the minerals which accompany it,—and particularly, when I had observed and considered all the circumstances attending its position or *repository*, or its relation to the rocks on which it rests, and, extending my views, had contemplated the basaltic mass as a whole ; —then all my ideas about their volcanic origin

vanished, and in the basalts of Saxony, or even of all Germany, I no longer perceived any thing but rocks formed after the same manner as all other mineral beds had been formed.

———————

ACCOUNT OF THE PROPERTIES OF BASALT.

(Added by the author subsequent to the reading of the Memoir before the Institute.)

I SHALL here give an account of the characters or properties of basalt, and of the kind of mountains which it forms, assuming that I may be allowed to found on the conclusion which I have already drawn (§ 32.) concerning its origin. I shall first consider the properties of basalt as a mineral substance, taken by itself, and shall then give a view of its geognostic characters.

The colour of basalt, as already remarked, § 57, is greyish-black; the black colour of the hornblende, which enters its composition, being more or less modified by the grey of the felspar. It is known that basaltic hornblende contains a good deal of iron ; according to the analysis of Lampadius, it includes also a little carbon : these are the two colouring principles. Klaproth likewise has found a little carbon in basalt : he says, " The black colour of basalt proceeds not only from the oxide of iron which it contains, but also, as has been well ascertained, from a small portion of carbon *." Occasionally the hornblende has a green colour, perhaps from a superabundance of magnesia; the basalt then has a tinct of the same colour ; and in this case it approaches sometimes to wacke, and sometimes to greenstone. The oxidation of the iron is the cause of the reddish or brownish colour observable in certain basalts, especially such as are vesicular.

I shall speak of the external shape, when I come to treat of the geognostic characters of basalt.

---

* *Journal des Mines*, No. 74.

The characters of its fracture are merely such as a clayey substance highly compacted, might be expected to present.

In hardness and difficult frangibility, basalt resembles all the other hornblende rocks. I have seen at Johann-Georgenstadt, one of these hornblende rocks, a sort of greenstone-slate, which was so difficultly frangible, that the people employed it in place of iron, for arming the head of the stamps of a stamping-mill. Each of these stamps weighed about 200 weight. The stones broken by this machine, were very hard; the greater part were of quartz. Little more than half an inch of this hornblende mass was worn down in the course of working for a month.

The specific gravity is a little under that of basaltic hornblende, which, according to Haüy, is 3,250; while that of basalt is, according to Klaproth, 3,065. Supposing basalt to be a compound of three parts basaltic hornblende, and one part felspar, laying aside all foreign ingredients, we would have for the specific gravity of basalt, $2,438 + 2,64 = 3,078$; a result which approaches very near to the statement of Klaproth.

The magnetic properties possessed by some sorts of basalt, proceed probably from a certain quantity of iron existing in them, nearly in a metallic state. In some basalts, the iron is distinguishable by the eye at first sight: it occurs under the form of small black and shining points, which, when they become enlarged, form those grains of magnetic ironstone, often to be observed in this mineral. These are sometimes so abundant, that the loose matter proceeding from the decomposition of the basalt, is gathered and smelted, as at Naples, in Virginia, and other places.

The greater or less facility with which basalt decomposes when exposed to the influence of the atmosphere, proceeds from a difference in composition. When it is very black, or when it is divided into prisms resembling in colour columns of iron, it appears completely to resist the action of the elements. In these cases, it is probably nearly a compact mass of basaltic hornblende. But the basalt which is less hard, and of a greyish colour, contains more felspar:

it is well known, that that substance is very liable to decompose, and its decomposition must involve the gradual destruction of those rocks of which it is a constituent. When the basaltic hornblende contains a greater proportion than usual of lime and of magnesia, or, in other words, when it approaches to actynolite, and forms what Werner calls Common Hornblende; the basalt rock composed of it, is of a greenish colour, and less hard; it approaches to the nature of wacke, and decomposes readily.

Exposed to the action of fire, it is converted into a black glass, if cooled quickly; and into a black substance of a stony aspect, when cooled very slowly. A mixture of grains of hornblende and felspar, that is to say, *greenstone*, affords the same results. According to Saussure, basaltic hornblende melts at 90–140 degrees of the pyrometer of Wedgwood; felspar at 70°; and basalt at 76°. If, while the basalt is melting, it be kept in contact with charcoal, the iron contained in it, is de-oxygenated; the melted mass no longer forms a black glass; and the revived metal is found in small globules.

I have elsewhere (Note A.) spoken of the
geognostic characters of basalt, and I shall now
give a more minute description of them.

58. The crystals which we find in basalt,
and which give it a porphyritic structure, have
been produced, during the precipitation of the
basaltic mass, by some elementary principles
uniting according to the laws of their affinities,
and forming the integrant particles of certain
minerals; the affinity existing between these
particles leading them to unite in the most
suitable manner, and thus to produce crystals.
The power of affinity in these particles, has in
many instances been sufficient to overcome the
resistance offered by intervening basaltic particles, and the crystalline form is complete.
But in other cases, this power has not been sufficient to turn aside all the basaltic particles,
but has operated only to a certain extent, so
as to produce a grain with a lamellar structure
in place of a crystal.   Further, I have often remarked, that these grains were of a form approaching to that of the crystal which would
have been produced, had no obstacle inter-

vened. The crystals, perfect or imperfect, found in basalt, must have been formed at the same time with it, or at least while the mass was soft, and before its parts had acquired a certain degree of coherence.

The crystals which occur in all rocks having a porphyritic structure, are, as has been already remarked, very uniformly or equally distributed through the mass of the rock; and they are isolated, it being a rare occurrence to find a few grouped or in contact. I think it is easy to assign a reason for these appearances. Almost all the integrant particles of crystals have united in the midst of a soft mass, in consequence of their chemical affinity: all those which happened to be within the same sphere of attraction, have united around one centre of action, or, in other words, have formed a crystal; and the number of crystals produced, has been exactly equal to the number of partial spheres of attraction.

59. In explaining the amygdaloidal structure, it seems necessary, first, to assign a cause for the vesicular cavities, and then to account

PROPERTIES OF BASALT. 177

for these being filled, in whole or in part, with the mineral substances which they contain.

It is probable that the vesicular structure has been occasioned merely by the disengagement of bubbles of some elastic fluid produced during the formation of the rock, the viscidity of the matter enabling it to retain the bubbles in the place where they had been formed. This elastic fluid, according to all appearances, has been carbonic acid gas, which had owed its origin to the carbon which we know to be contained in basalt, united to the oxygen of the oxide of iron. I feel some difficulty, I must confess, in conceiving how such a cause could so affect a bed of basalt, perhaps 150 feet in thickness, as that it should, in nearly its entire extent, exhibit a vast number of vesicular cavities, equally dispersed throughout the mass. The fact, however, does exist; and we may perceive something analogous in several other mineral substances; for example, in some ores of manganese,—in the masses of certain veins, such as those of Graupen in Bohemia, or of Annaberg in Saxony, (§ 22.)—in the *toadstone* of Derbyshire, (§ 42.)—but parti-

M

cularly in the bog iron-ores, which may be said
to be forming under our eye in the marshes of
Silesia and Lusatia. This vesicular property is
perhaps peculiar to clayey substances, impregnated with oxide of iron, and containing a
small portion of carbonaceous matter : basalt is
of this nature.

After the formation of these rocks containing vesicular cavities, water may have
oozed through them, particularly in the early
period of their existence; and it would naturally deposite in these cavities the substances
which it held in solution : there would thus
be formed a first coating or layer on the walls
of the cavities, and a number of successive additional layers would necessarily fill them. Many
of the cavities are not completely filled ; these
constitute *geodes,* and their interior is sometimes studded with crystals. The greater part
of the substances which I have observed occupying the air-holes in basalt, are such as seem to
be most under the influence of water ; as, green-
earth, or earth of Verona *, steatite, calcareous

---

* So called from its abounding in the amygdaloid of
Monte Baldo near Verona.

spar, zeolite; although it is true that calcedony
and quartz also occur.

The kernels, geodes, and grains which belong to the *amygdaloidal* structure, must not be
confounded with the grains which constitute
the *porphyritic* structure.   The former have been
posterior in formation to the basalt in which
they are found; the latter, contemporaneous.
In the porphyritic structure, the grains always
exactly and completely fill the cells which they
occupy, their shape is generally angular, and
they are solid.   In the amygdaloidal structure,
on the contrary, the form of the grains is either
spherical, or at least rounded ; the cells are often empty ; sometimes the walls are merely
lined with a simple coating ; and in other instances, the grains themselves are hollow.

Lest the density or compactness of basalt
should seem to be a reason for doubting if a
fluid could pass through a mass of it; I shall
mention a striking instance which fell under
my own observation. At Almerode, near to
the Meisner mountain in Hessia, there was formerly a lead-foundery, the furnaces of which
were partly constructed of basalt, containing a

great number of globular cavities. When these
furnaces afterwards came to be demolished, and
the stones broken, many of the interior vesicles were found to be filled with lead and its
oxide. In the collection of the director of the
salt-works of Ahlendorff, I saw a curious specimen of the basalt from these furnaces : the cavities were about half an inch in diameter ; some
were empty ; some were full of minium ; others
contained balls of lead, the surface of which was
covered with a thin coating of yellow litharge ;
and the vesicles in one corner of the specimen
had their upper hemisphere sprinkled with
white oxide of lead, forming a sort of downi-
ness, shooting into minute plumous crystals.
After all, it must be admitted, that the heat of
the furnaces may possibly have dilated the pores
of the basalt, and thus have rendered it more
permeable to the melted metal.

60. The divisions observed in basaltic masses,
are, 1. Into columns ; 2. Into tabular masses ; 3.
Into balls composed of concentric layers ; and, 4.
Into what are called granular distinct concretions.

The division into *columns* or *prisms*, appears
to have been the effect of a contraction which
took place in the basaltic mass, in the course
of drying and hardening. All clayey masses,
in drying, shew a tendency to this sort of division. I shall not repeat any thing that I have
already said on this subject, in § 27. The contraction produced cracks, which, inclining to
assume two, three, four, or more directions,
very generally parallel to the same plane, divided the mass into prisms with four, six, or
eight sides. These fissures are not straight,
but are subject to a number of bendings. Sometimes the divisions extend only a short way into the basaltic mass; from this cause, proceed
many irregularities in the columns. Yet even
in this sort of division, there is an evident tendency to a certain regularity. I have generally
observed, that the more homogeneous, compact
and hard the basalt, the more did the columns
approach to the figure of regular hexagons.
At Stolpen, there are places where the columns
have as much regularity as artificial prisms of
cast iron could possess. In one part of the
court of the castle, as formerly mentioned,

which has been formed by cutting across a
group of perpendicular columns, the natural
surface has exactly the appearance of a pavement of hexagonal stones. It would seem as
if six were the centre or medium of the varia-
tion as to the number of sides assumed by
columns of basalt; the cause of which remains
a problem.

The division into *tabular masses* is also to be
considered as the effect of contraction; the fissures in this case having run only in one direction. I have already, (§ 15.), stated sufficient
reasons for not considering this tabular division
as a kind of stratification.

61. Blocks of basalt often appear in the form
of *balls*, surrounded by concentric layers like a
sort of bark. Both these appearances, I am inclined to ascribe to the influence of the atmospheric elements. I may, in the first place, remark, that I have never observed such balls
with concentric layers, any where but among
the fragments scattered along the bases of basaltic mountains, or in the neighbouring fields :
unquestionably, they are never to be seen among

the unaltered rocks in basalt-quarries, where the atmospheric influences have been excluded. These coated balls might, I think, be produced in the following manner,

Basaltic columns are frequently traversed, at different distances, by fissures perpendicular to their axes, (§ 8, 9). If one of these columns happen to fall off, and roll to the bottom of the mountain, it will break in the line of the fissures, and will thus become divided into a number of angular fragments or blocks, the three dimensions of which will often be nearly equal: the angles and edges will soon suffer decomposition, being the parts most exposed to this wasting influence; and the block will come to assume a rounded form. The action of the elements will proceed towards the interior; the outer part, which must be first affected, will undergo a change of density, probably a dilatation or expansion; by this means, it will become separated from the rest of the ball; somewhat in the same way, as occasionally happens in a piece of glass, or of earthenware; when one portion only suffers contraction or expansion, it separates from the rest by a crack. The decomposition always proceeding farther towards the interior, at each step (if I may so express myself) detaches a new coating, or adds a new layer to the bark. When one of these balls is broken, the centre or nucleus is found to be black and compact; the layers which surround it, form a sort of bark, perhaps two inches in thickness: the individual layers are of various thickness, and they are more grey and more earthy the nearer they are situated to the surface.

It may perhaps appear extraordinary to some persons, that the action of the atmospheric elements should penetrate to the interior of a mineral mass, without producing the absolute destruction of the exterior parts in its progress. But I shall mention a convincing instance of the possibility of this, observed in basalt itself. I have often seen balls of that substance, perhaps from a foot to sixteen inches in diameter, which had been recently broken. The centre was fresh, black, and hard; it was surrounded with concentric layers of a greyish, almost earthy substance, appearing like a sort of bark, from an inch to an inch and a-half

thick. The decomposition did not seem to
have proceeded any farther; for within this
bark-like substance, the basalt did not appear
to be altered. But the olivine which it contained, bore evident marks of partial decomposition, to the distance of three, or three and a
half inches within the concentric layers. The
grains of this substance, which had formerly
existed on the surface of the block of basalt,
were completely gone, leaving only empty
spaces : those next in order, proceeding inwards,
were also entirely decomposed, and converted
into an earthy substance : the next again were
an assemblage of grains without coherence :
then the colour only was found to be altered :
and at last, at the distance of about four inches
from the surface, the olivine still retained both
its original solidity, and all the freshness of its
colour.

As to the division into *granular distinct concretions,* observable in a great number of basalts, it
appears to me to be an effect of the original formation, rather than of any contraction which
might happen in drying. Possibly these two
causes may have concurred in producing the

effect. While the deposition of basaltic sediment was gradually going on, certain particles
uniting many others around them, seem in some
sort to have formed particular bodies in the
middle of the general mass.

62. I have already (§ 29, 30, 47, and Note A)
mentioned the relations of basalt with the other
rocks of the same formation, (that of the *traps,*)
or, in other words, of the same family, as they
all owe their origin to the same solution. This
solution had contained chiefly the constituent
principles of hornblende and felspar. While
it was in a state of tranquillity and purity favourable to crystallization, these principles by
their union in a crystalline form, would produce
greenstone. If the constituent principles of felspar happened to be in greatest abundance in
the solution, then there would be formed either
a *greenstone,* or more frequently a *porphyry-
slate,* according as the precipitation was more
or less troubled : but if the constituents of
hornblende prevailed, as they generally do,
then basalt would be the result. Again, if the
precipitation had been still more impure,

and had become coarser, the sediment would form wacke, and at length clay. I need scarcely say, that I can only offer these as conjectures founded on probability : but they do appear to me to possess all the presumptive evidence to be expected concerning facts of this description in geology. In the first place, there can be no doubt that these substances belong to one and the same formation ; they invariably occur together, and they pass into each other by insensible gradations. Further, the results of chemical analyses confirm what I have stated concerning the differences in composition : the table introduced into Note H. shews the chemical composition of Basalt ; and the following exhibits that of Porphyry-slate :

	Alloy or mixture of ¾ felspar, and ¼ basaltic hornblende.	Analysis of porphyry-slate by Klaproth.
Silica, ...	56.38	57.25
Alumina, ...	19.50	23.50
Iron, (oxide), ...	7.00	3.25
Lime, ...	3.50	2.75
Manganese, (oxide), ...	0.00	0.25
Alkali, ...	9.75	8.10
Water, ...	0.00	3.00
Loss, ...	3.12	1.90

I once more repeat that I do not pretend to draw any positive or definite conclusions from these comparisons; but submit them only as affording hints. And certainly if the alkali found in felspar be uniformly potass, and that found in trap-rocks be soda, the conjectures now thrown out must be abandoned.

63. In regard to the relation of basalt with rocks of other formations, we find that it constantly lies over them, and is never itself covered. We are now to take a cursory view of the manner of its overlying other rocks. It has already been stated, (§ 26. and Note A) that the *surface of superposition* covers the ends or *outgoings* of the mineral-beds on which it rests, and is not parallel with those beds; from whence we concluded, that the solution which produced the basaltic mass, must have stood at a higher level than these, and that the rocks on which the basalt was deposited, had previously experienced some wasting or disintegration. The many hills which are composed of grey-wacke, sandstone, breccia, or puddingstone, and the large collections of gravel, of sand, and

of clay, are so many proofs of former disintegrations, since they are nothing else than the
detritus of preceding rocks.   Many of these
hills are of more ancient date than basaltic
rocks, being covered by them; there must
therefore have been a wasting and disintegration of rocks anterior to the existence of any
basalt whatever.

Viewed on the great scale, basalt occurs in
patches separated from each other, and it forms
mountains of a conical shape.   Let us consider these two facts somewhat more closely.

In Note I., are detailed certain causes which
might have destroyed the continuity of the
great basaltic mass, and divided it into scattered portions.   For my present purpose, it is
enough to state, that we have the most incon-
testible proofs of the disintegration and destruction of rocks, and that these could not be attacked and wasted, till the covering of basalt,
which protected them in many places, was destroyed.   If some portions of the basaltic mass
have remained firm and untouched, in the midst
of the destruction of those which surrounded

them, the cause is to be found in a difference
in the hardness, and perhaps also in the nature
of the basalt.   I have already mentioned instances (§ 8, & 29,) of such a difference in
hardness between neighbouring basalts, and
even between different parts of the same basaltic mass ; and an additional example is given
in Note U.   The black hard basalt, which is
divided into columns resembling pillars of iron,
seems effectually to resist decomposition : it
may perhaps be considered as little else than
compact basaltic-hornblende : but that kind of
basalt, which contains more felspar, lime, and
magnesia, and seems thus to approach to wacke,
easily softens and decomposes.   This sort would
yield to the destructive action of the elements,
and would be washed away ; while the other
would remain unaffected : it may therefore be
the only kind now in existence in many places.
In this way, we may account for the broken
and shattered aspect of the great basaltic mass
which covers a part of the globe.

64. When the softer portions of the basalt
were once destroyed, the rocks which they co-

vered would become a prey to the destructive
action of the atmosphere : their surface would
become lower and lower in level, especially if
they happened to be situated on one of the elevated parts of the globe, near the ridge of a
chain of mountains : while the hard and in-
destructible portions (if I may use so strong
an epithet) would remain at their original level,
supported and maintained at that height by the
subjacent rocks, which, by their means, had
been saved from disintegration. These rocky
portions, thus preserved, would form a number
of isolated mountains, the summits of which
would consist of basaltic platforms. The vertical falling of rains, the continual decomposing
action of the elements, and the effect of gravity
on a friable mass such as that of rocks, must
necessarily produce the conical shape. It seems
needless to insist farther; as a little reflexion
must produce conviction. Supposing that the
atmosphere could have little effect on the basalt of the platforms which form the highest
part of these mountains, some mechanical
causes, such as thunder, frost, and in certain
places even the hand of man, may have contri-

buted to the detaching and throwing down of basaltic columns. Such disintegrations would necessarily affect the higher part of the basaltic platform more than the lower; so that the platform, in diminishing in size, would assume a rounded, or even conical form ; and mountains thus furnished with basaltic summits, would no longer be trunks of a cone, but complete cones.

Basaltic mountains are generally thickly clothed with vegetation, often with fine trees. This arises from the rich quality of the earth which is produced by the decomposition of basalt, and perhaps also from the humidity incident to basaltic mountains. Springs very generally appear at the base of such mountains which we may perhaps explain, by supposing, that the basalt, by reason of its density, and possessing a considerable power of conducting caloric, fixes and condenses the vapours of the atmosphere. These properties render basalt unfit for a building stone ; for, in winter, the walls of houses built with it, are found to be damp within.

NOTES,

CHIEFLY BY THE AUTHOR.

~~~~~~~~~~~

(The few Notes added by the Translator, will be found properly distinguished.)

NOTES, &c.

―――――

NOTE A.   (Page 15, &c.)

It may be useful to state the circumstances which Professor Werner endeavours to determine concerning a rock, when he considers it in its geognostic relations. These circumstances are, 1. *Its structure in the small,* which can be ascertained by cabinet specimens : 2. *Its structure in the great,* which can be seen only in large masses, or in entire mountains : 3. *Its relation to other rocks of the same* FORMATION, (or those rocks which generally accompany it) : 4. *The situation of the whole Formation with respect to other Formations of rocks,* including position and stratification.

I shall attempt very briefly to illustrate these topics, referring for details to Werner's " Elementary Course of Geognosy."

1. The following tabular view of the structure of rocks, will sufficiently explain what is meant by the expression *structure in the small* of a rock, and will at the same time shew its different kinds.   Werner calls this sort of structure, *Structur des Gebürgs-gestein,* literally the " structure of the stone of the rock."

TABLE.

STRUCTURE				
Simple,			....	Limestone, quartz, clay-slate, &c.
Compound,	Cemented,	Irregularly,	....	Sandstone, puddingstone, breccias.
				Verdé antico.
	Aggregated,	Simply aggregated,	Granular structure,	Granite, sienite, greenstone.
			Slaty structure,	Mica-slate.
	Regularly,	With a basis,	Porphyritic structure,	Porphyries, some basalts, &c.
			Amygdaloidal structure,	Wackes, toadstone, &c.
Doubly compound,	Granular-slaty structure,		....	Gneiss, topaz-rock *.
	Granular-porphyritic structure,		....	Many granites, &c.
	Slaty-porphyritic structure,		....	Mica-slate with garnets, &c.
	Porphyritic-amygdaloidal † structure,		....	Some basalts, &c.

* In the tabular view of the structure of mountain-rocks given by Professor Jameson in the third volume of his System of Mineralogy, an additional kind of the doubly compound and aggregated structure is introduced, in order to embrace the Saxon topaz-rock, which is not *granular-slaty*, or " slaty in the great, and granular in the small," but *slaty-granular*, or " slaty in the small and granular in the great."

T.

† In the same table, Professor Jameson denominates this kind of structure, the " porphyritic *and* amygdaloidal ;" observing, that in the granular-slaty, slaty-granular, granular-porphyritic, and slaty-porphyritic, one structure is comprehended in another, or a smaller in a greater ; but that in the first mentioned, one structure does not include the other, but they are merely placed near or beside each other.

It may be added, that when speaking of *regular* or *irregular aggregation*, Mr Jameson prefers the epithets *determinate* and *indeterminate*.

T.

2. Mineral masses, considered in regard to their *structure in the great*, are composed of different strata placed one above another; or they are formed by a collection of columns, tabular masses, &c.; or else they consist of a continuous and solid mass. The circumstances attending the stratification of mountains composed of beds, ought always to attract the particular attention of the geognost.

3. There are several different kinds of rocks, or of mineral substances, which are almost always found together, and which thus appear to have been formed at the same period. Such rocks are said, in the language of Werner, to be *of the same formation.* Thus, certain kinds of sandstone and of slate-clay, very generally accompany coal, and alternate with it in beds : such sandstone and slate-clay belong to the Coal Formation.

4. When one formation of rocks is to be compared with another in regard to position *(gissement)*, it is essential to determine which of the two is really placed over the other : by this means, their *relative age* is ascertained : for it is evident, that any particular mineral bed must have been formed posterior to another which it covers. Thus we know, that the *formation of gneiss* is newer than that of granite, because we find the former of these two rocks uniformly placed over the latter. The next object is to determine whether the edge or superior extremity of the upper bed, is at a lower or higher level than that of the other; and whether the *plane of separation (surface de superposition)* is parallel to the seams of stratification of the mountain in which the bed is situated. From the correct observation of these particulars, which it is often difficult to

accomplish, very interesting conclusions are to be drawn concerning the formation of rocks.

Let us carry back our ideas to the time when the present crust of the globe was formed. The nucleus must of course have previously existed. It was surrounded, we may suppose, with an universal ocean, holding various matters in solution : the precipitates or sediments were deposited one above another : each of them formed a bed or a stratum, which, in moulding itself over the preceding, naturally followed the bendings, heights and hollows which occurred, maintaining throughout nearly a parallel thickness. The seams which separate the beds or the strata, are called *seams of stratification.* At that time, the lower surface of each bed or stratum, which is called the *surface of superposition,* or *plane of separation,* must have been exactly parallel to the seams of stratification ; and as all the layers must then have surrounded the whole globe, they could have no upper edges or *outgoings.*

Let it next be supposed, that the waters had become lower in their level, so that certain elevated points or mountains remained uncovered ; it is evident, that no subsequent precipitates or sediments could form beds surrounding the whole globe : their upper edges would necessarily be wrapped around the elevated points alluded to, which would afterwards appear as if they had pierced those beds, and risen above them. In proportion as the waters became lower in level, the *outgoings* or upper edges of the beds successively formed, would necessarily be lower : but the line of separation of each bed, (as well as the general surface formed by their assemblage), would constantly continue parallel to the seams of the stratification.

Let us further suppose, that when the waters had diminished to a certain level, a disintegration or destruction of a part of the solid matter already deposited took place ; it is evident, that if such disintegration affected one part of a mountain more than another, the surface of the mountain would no longer maintain its parallelism with the seams of stratification. If, at this time, the waters should again rise, and the older and denuded part of the mountain should be covered with a new formation of beds, the seams of stratification in this formation would be parallel to the surface, but not to the stratification of the beds on which they rest. Now, the *surface* of these beds is the *surface of superposition* or plane of separation of the new formation. Thus, whenever we find a formation of mineral beds, the plane of separation of which is not parallel to the stratification of the rocks on which they rest, we may safely conclude, that such rocks have undergone a certain degree of wasting and disintegration before the new formation took place ; or, what appears generally to have been the case, that a long interval elapsed between the periods of the two formations. In the same way, if the *outgoings* of beds of clay-slate be found uniformly lower than those of beds of gneiss, we may conclude that the solution when it deposited the former of these two substances, was at a level inferior to that at which it stood during the deposition of the latter. But if a mineral mass is incumbent on another, and consequently more recent, and at the same time covers the edges or *outgoings* of the inferior mass, it is evident, either that the uppermost mass must have owed its origin to a new solution, which had risen above the

N 4

level of the former, or that the first solution itself had again risen to a higher level.

We shall now apply this method of considering rocks, to the particular instance of basalt.

(1.) The *structure in the small* of basalt is sometimes *simple;* but more generally it is *porphyritic;* and not unfrequently it is at once *porphyritic and amygdaloidal.* It is generally *porphyritic;* because in the midst of the basaltic mass or paste, there occur crystals, or grains (which may be considered as imperfect crystals), of basaltic-hornblende, of olivine, augite, magnetic-iron-stone, mica, leucite, felspar, and melanite. These crystals or grains have been formed at the same time with the basaltic mass; they completely fill the little cell which they occupy, and adhere to its walls; they are pretty equally distributed through the mass, and scarcely ever grouped. The structure of certain basalts is *amygdaloidal;* because small vesicles or cavities occur in the mass, partly tortuous, and partly spherical; sometimes in such numbers, that the mass resembles a sponge. These vesicles are sometimes empty; and sometimes occupied, in whole or in part, by foreign substances, which have probably entered by infiltration: in some cases these substances line the walls of the vesicles, with a simple coating; in others, various layers, often differing in nature, are deposited one over the other, and form geodes and agates; and in other instances still, they entirely fill up the cavities. The foreign matters which are most commonly found in basalt, are steatite, green-earth, calcareous spar, zeolite, calcedony, and quartz.

(2.) Basalt, viewed as existing in great masses, is found divided into beds, which again are subdivided

into columnar and tabular. The form of the columns
is more or less regular; and I believe it will be found,
that, in proportion as the basalt is compact and fine
grained, the column approaches to a regular hexagon; but I have seen such columns with three, four,
five, seven, and eight lateral faces. In other respects,
these polyhedra have nothing in common with those
which are the immediate product of crystallization:
their edges are more or less blunted: the inclination of
the sides towards each other, is by no means uniform;
sometimes it is 120 degrees at one extremity, and 100
at the other; sometimes there are six sides at one end,
and but five at the other: the lateral faces are not always planes, but frequently present concavities and
convexities, which join into the convexities and concavities of the adjacent faces of the contiguous prisms.
These columns have generally a direction approaching to vertical; sometimes, however, they are considerably inclined; and even entirely horizontal. Very
generally they are collected into distinct groups, each
group having a particular position: I have observed
some, (§ 14.), in which the columns diverged from a
centre like the rays of a sphere; and in the *Journal de
Physique, messidor an* 10, I have described a group
which pretty accurately represented an immense palm-
leaf. The columns are often divided at regular intervals, by fissures perpendicular to the axis, and which
are often curved or concave; the prisms thus assuming
an articulated appearance. When such columns happen to fall down and break, the fragments, worn by
the destructive agency of the elements, assume the form
of balls, composed, at least in the exterior part, of concentric layers. I have mentioned a striking example

of this sort of division, in the *Journal de Physique,* *messidor an* 10, and have explained its production at § 61. of this Memoir. Basalt also occurs divided by parallel seams into plates or tables, which vary from half an inch to several inches in thickness.

(3.) The beds or masses of basalt are often accompanied by beds or masses of wacke, greenstone, or porphyry-slate. These mineral substances, taken together, compose mountains ; they are only found in the neighbourhood of each other ; and they gradually pass into each other, by a progressive series of characters. It follows, that they are of the *same formation.* The wacke occurs in thin beds under the basalt, and the greenstone generally above it ; so that the basalt passes, on the one hand, into wacke, and on the other into greenstone. In the former case, the basaltic precipitate or sediment passes into a mechanical or earthy sediment ; in the latter, into a chemical precipitate, or crystalline form. The porphyry-slate is found in the vicinity of basalt, and appears to be of a more crystalline nature. The conclusions to be drawn from these facts have been already stated, § 29, 30, 37, 47, Note K. ; and for an explanation of them, I refer to § 62.

(4.) Considered in relation to rocks of a different, formation, it may be affirmed that basalt very generally lies over them, and is never covered by them : it follows, that the *formation* of basalt is newer. We have here to attend only to the particular circumstances of its *superposition.* Werner, in his geognostic language, which is equally concise as expressive, speaks of basalt resting on other rocks in an *overlying,* an *unconformable,* or a *broken* position, *(ubergreiffende, ab-*

weichende, unterbrochene Lagerung), that is to say, the basalt covers the head or ends of the beds,—the plane of separation between the basalt and the other rocks, is not parallel,—or, taken as a whole, the bed of basalt does not form a continuous mass, but is divided or broken.

Lastly, basalt generally occurs on the summits of conical and isolated mountains.

~~~~~~~~~~~~~~~

Note B. (p. 20.)

Pseudo-volcano at Epterode.

In Hessia, about a league from the Meisner mountain, in the direction of Cassel, and near the village of Epterode, there is a small hill composed chiefly of scorched earth and stones. It has the appearance of a great collection of refuse from ancient brick-works. Among the substances composing the heap, I remarked porcelain-jaspers, earths scorched and roasted, ferruginous scoriæ, some substances completely vitrified, fragments of quartzy-sandstone, some parts of the surface of which bore unequivocal marks of fusion, and even pieces of basalt, partly scorified, and agglutinated to fragments of porcelain-jasper. In one place, if I be not deceived, I saw the mineral beds still in their original position; they were of a red colour, much scorched, but not vitrified. All these facts indicate a *pseudo-volcano;* that is to say, a bed of coal has been burning in that place, and has calcined and

vitrified the different substances which lay within the
sphere of activity of such a subterranean fire. Almost
all the tops of the neighbouring hills are covered with
basalt, under which, there sometimes occurs a bed of
coal, accompanied with sandstone and clay. At the
Meisner mountain, the coal is at present on fire : this
had formerly been the case at Epterode, which is one
of the lower basaltic summits.

I have described another pseudo-volcano in the *Journal de Physique, messidor an* 10.

NOTE C. (p. 21.)

Description of the Meisner mountain.

IN the midst of a mountainous country, at the distance
of eighteen miles to the eastward of Hesse-Cassel, the
Meisner rises like a Colossus above the surrounding
hills, from which it stands entirely separated. The
summit of this huge mass forms a plain, about six miles
in length, and three in breadth. The mountain is
partly naked, and partly covered with wood. Its sides
have a rapid declivity. The whole surface-rock of
this part of Hessia, consists of shelly-limestone, alternating, in some places, with a red sandstone ; upon
which, there occasionally lies a shistose-sandstone, (or
sandstone-flag), which contains much mica, and decomposes very readily. Standing to the eastward of the
Meisner, the direction of the general stratification of

the beds, appears to be between east and south: the inclination is sometimes very considerable.

The body of the mountain consists of limestone and red sandstone: these substances constitute the mass, to the extent of three-fourths or of five-sixths of the height, calculating from the western base. I had not time to determine the exact order of their arrangement. I observed in the limestone, beds of clay and of gypsum. The body of the mountain is covered by a bed of the sandstone-flag above mentioned, which is friable, and in some places impregnated with bitumen: it contains also some collections of sand: this bed is nearly horizontal, inclining a little, however, towards the south-east, especially on the east side. Over this lies a great bed of bituminous substances: its thickness varies considerably; in some places, it is not less than ninety feet thick; but more commonly it does not exceed twenty feet; often sixteen feet, and even twelve; and at some points I observed it become thinner and thinner, till it at last disappears, the *roof* joining to the *floor*. The lower portion of this bed consists of bituminous wood *, of a yellowish-brown colour, with the woody texture in entire preservation, and of bituminous earth, or brown coal, (Braunkohle). In tracing higher up the bed, the quantity of bitumen seems to increase; towards the top of it, there occurs a pitchy substance †, or unctuous jet ‡; and the uppermost portion generally consists of a singular

---

* Brochant's Mineralogy, vol. ii. p. 44, and 492.

† Brochant, vol. ii. p. 49.

‡ Pitch-coal of Jameson.

variety of bituminous matter, divided into small prismatic columns, (bacillaire or Stangenkohle, the columnar coal of Jameson *.) This mass is almost every
where covered with a bed of clay, which is generally
of a black colour, and impregnated with a good deal of
bitumen. The abundance of bituminous matters here,
has given rise to many mining operations. These substances very easily catch fire; and for ages past, several parts of the mass have been burning.

Upon this mass of coal, or on the clay which accompanies it, rests the enormous basaltic platform, which
forms the summit of the mountain. This platform
varies in thickness, from 300 feet to above 600 feet;
and in length and breadth, it extends entirely across
the upper part of the mountain. Its lower surface
seems to follow the general stratification, and makes
some bendings along with the other beds; and I can
affirm, that it is nearly horizontal, at least to a very
considerable extent; for the level belonging to the
works, is more than 1600 fathoms in length; and for
at least 1000 fathoms of this extent, the level is nearly horizontal, and is chiefly cut in the bed which has
the basalt for its roof. I have, besides, explored many
transverse galleries also excavated in the same bed;
all of which had only a slight inclination. The upper
surface of the platform likewise, has only a very small
inclination; it is indeed almost a level. At the same
time, it presents, on the east side, a wide defile or
ravine, which begins towards the middle of the platform. This ravine has probably been produced by the

---

* Brochant, p. 51.

long-continued action of the waters which run in that direction; or, it is very possible, that such a hollow may have existed in the body of the mountain when the basalt was deposited, and that this substance may have moulded itself into it. The vulcanists find in this appearance the vestiges of a crater.

In examining the circumference of the platform, many breaks or chasms occur, which exhibit its interior structure. In one of these, towards the west side, the basaltic mass is divided by fissures into columns, chiefly six-sided, but very irregular: they are from three to six feet long, and about four inches in thickness. The lateral faces present a succession of convexities and concavities, which join with the concavities and convexities of the adjacent faces of the contiguous columns. They are arranged in horizontal beds; and have the most striking resemblance to logs of wood piled in a timber-yard.

The basalt is very black, and very hard and compact. It contains, at distant intervals, some minute grains of olivine of a very fresh colour; but more commonly it exhibits small vesicular cavities, filled with a sort of greenish steatite. The upper part of the platform seems to be almost entirely a greenstone, (called *Dukstein* by the people of the place), or a rock composed of grains of hornblende and felspar, the former ingredient being the more abundant. This bed of greenstone is mentioned by Werner in his Theory of Veins, § 49.*; it is described by M. Schaub of Cassel,

---

* " Greenstone occurs on the Kolbe of Weissner, very well marked: it consists of hornblende in large grains, mixed with much felspar well characterized: it is there called *Dukstein*."—*Dr Anderson's translation*, p. 76.　　T.

in his book on the Meisner; and that eminent mineralogist and writer M. Voigt, who has exerted great
talents in maintaining the vulcanic origin of basalt
against Werner and his pupils, thus speaks of it, in his
Mineralogical Travels in Hessia, 1802 : " The rocky
mass, called by the inhabitants Rebbes, (or the Milk-
pot), as well as several other rocks which project
through the turf in that part of the mountain, are
composed chiefly of a mixture of hornblende and felspar, (though I cannot answer for this last substance
being a true felspar) ; to which mixture, Werner has
given the name of *grunstein*."

The Meisner greenstone is in most places nearly
compact, so that it is difficult to discern the grains;
but in particular portions of the rock, the grains are
larger than peas. I have already noticed, § 30. the
gradual and complete transition from greenstone
with a granular structure, to the most compact basalt.

Several observers have considered the Meisner as an
ancient volcano, and the basaltic platform which covers
it, as lava. For my own part, although I examined
the exterior of the mountain, and also the interior as
far as practicable, with the most scrupulous exactness,
during the space of two days, I could perceive nothing
volcanic about it ; and I think I may assert with confidence, that it cannot be volcanic, for the following
reasons :

1*st*, There is here a perfect transition from the most
compact basalt to greenstone with a granular structure : these two substances are so generally interwoven,
and blended together in so many places, and they pass
so evidently into one another, that it is impossible not

to perceive that they are only modifications of one
and the same substance, and that both have had a similar origin. Now, the granular part, far from owing
its origin to fire, cannot even have undergone the action of that element; for it does not bear the slightest
mark of alteration: the grains of hornblende and felspar have preserved their lamellar structure, their
lustre, and their freshness; the compact part, therefore, or the basalt, which is intermixed with the granular, cannot be a lava. The same sort of argument, in
a similar case, produced conviction in Dolomieu.
(See § 30.)

2dly, A volcanic mountain is composed of a great
collection of fragments, stones, scoriæ, cinders, and
heaps of melted substances, accumulated without any
order: it seems evident, therefore, that a mountain
composed for five-sixths of its height of shelly-limestone regularly stratified,—this covered by several beds
of sandstone, and a bed of coal,—and the whole sur-
mounted by a continuous mass of rock, partly compact
and partly granular in its structure, cannot be a volcanic mountain.

3dly, A volcano throws out lavas, which, like all
other fluid substances, obey the laws of hydraulics, and
form a sort of stream descending to the base of the
mountain: but a volcano cannot certainly be supposed
to have emitted a vast mass of viscid lava, and to have
fixed it on the top of a mountain, forming in that
elevated situation a platform nine miles long and three
miles broad, and more than three hundred feet thick.

4thly, A heap of melted stones so enormous as to constitute such a platform, could not be supposed to spread
itself over a bed of bituminous matters, without pro-

o

ducing some degree of alteration. In the mines of this mountain, however, I have in two different places seen the basalt in immediate contact with columnar coal, while that highly bituminous substance remains in precisely the same state as when covered by a bed of clay, which is most commonly the case.—I repeat the question, if it is possible to conceive that a great mass of melted stones spreading over a bed of bituminous and inflammable matters, should not affect them otherwise than would a simple sediment of clay ?

5thly, Although the transition from greenstone to basalt is not only completely proved, but is obvious, (§ 30), yet some may wish to deny it, in order to evade the consequences : this can avail them little ; for they must admit, that the basalt forms a bed situated between the coal and the greenstone which forms the upper part of the platform ; in other words, that the basalt occurs between a bituminous substance and a rock having a granular and crystalline structure ; —a situation which no vulcanist will choose to assign to his lava.

NOTE D.—(p. 31.)

Passage of Clay into Wacke, and of Wacke into Basalt.

PROFESSOR WERNER's account of these transitions as observed by him at the Scheibenberg mountain, is so interesting, that it deserves to be inserted entire.

" The unlooked for discovery, which I made last summer at Scheibenberg," (says that eminent mineralogist, writing in the year 1788), " concerning the relation of basalt to the subjacent rock, must, in my opinion, be extremely interesting to all impartial geognosts, especially at a time when the question about the nature and origin of basalt is agitated anew.

" In passing the Scheibenberg a good many years ago, I observed at a distance, and nearly at the top of the mountain, a heap of white soil. Having inquired what it was, I was told that in that place there was a sand-pit, or rather sand-mine, from which the people in the neighbourhood took sand for building. Having afterwards reflected that a sand-pit situated near the top of a basaltic mountain was an uncommon and curious occurrence, I resolved to inspect it more narrowly ; and, accordingly, accompanied by several of my pupils, I set out on this little mineralogical excursion.

" When still at some distance from the mountain I observed near its summit, a pretty deep rut or *gully*. I of course expected, that at this place the rocks would be laid bare, and the structure of the interior seen. At the same time, I had no expectation of finding any thing else than a bed of sand surrounding the foot of the basaltic summit, such as is generally believed to occur at Pœhlberg near Annaberg *. Great was my surprize, therefore, when, on arriving at the spot, the

o 2

---

* The subsequent observations of Werner, as well as the facts stated in § 9., (suprà, pp. 34, 35.,) shew that the bed of sand passes *under* the basalt at Pœhlberg.

first glance presented me with a regular stratification:
—Over a thick bed of *quartzoze-sand*, were placed several layers of *clay;* above these was a bed of *wacke*,
on which the *basalt* rested.   I was delighted to observe
that these three different beds extended nearly in a
horizontal direction, under the basalt.   In the lowest
bed, the sand became finer and finer upwards, till at
the top it was argillaceous, and then passed completely
into clay; in the same way, in the upper part of the
second bed, the clay gradually changed into wacke;
and this substance, again, passed into basalt.   In short,
there was here a complete transition from the purest
sand to clayey sand; thence to sandy clay; and, by
gradual shades of change, to fine clay,—to wacke,—
and at length to basalt.

" The striking scene before me seemed to justify such
conclusions as the following, which I almost involuntari-
ly pronounced at the moment:—The basalt, wacke,
clay, and sand, are all portions of one formation; they
are precipitates or sediments from an aqueous solution,
which had formerly covered the whole country; the
waters had at first brought *sand* into this place; then
they had deposited *clay;* the sediment gradually chan-
ging its nature, had given birth to *wacke*, and at length
produced true *basalt*.

" To these reflections, I have only a few remarks to
add.   The basalt exposed in the gully alluded to, was
divided into prisms, nearly vertical in position, and
distinctly separated from each other.   The prismatic
division not only continued down to the bed of wacke,
but extended some way into it.   The texture of the
wacke was slaty in the great.   It was not easy to
see the lower part of the bed of sand, as it was covered

by a heap of loose matter, dug from the mine: but the sand evidently became coarser and coarser in descending, till at length it degenerated into mere gravel. The gneiss, which constitutes the mass of the mountain all around, makes its appearance immediately below the heap above mentioned.

" I know not how those mineralogists who argue for the volcanic origin of basalt, will explain the transitions now described. For my own part, I am fully convinced, that all basalts have been produced in the humid way, and that they are comparatively of a very recent formation. I am of opinion, that in former ages they all constituted one vast continuous bed, which covered indifferently primitive and secondary rocks; that in the lapse of time a great part of this bed has been worn down and carried away; and that the numerous basaltic summits of our present hills are the mere shreds and remnants of the grand basaltic deposition.

Note E.—(p. 56.)

Mountain of Diberschaar.

This lofty and interesting mountain, is situated in the county of Glatz, at the distance of little more than a mile from Landeck. Its body is composed of gneiss and mica-slate, and it is capped with basalt. It is fully described by Mr Von Buch in his Mineralogical Description of the Environs of Landeck,—one of the best geological pieces which has appeared *. I can bear the most ample testimony to the accuracy of the account of Diberschaar, having myself seen all the appearances mentioned. At the top of the mountain, the basalt presents small cavities in such numbers, that it looks like a sponge; and certainly if I had seen a piece of this vesicular basalt lying near to some furnace, I might have taken it for a scoria: at the same time it has nothing of the vitreous appearance, but must have contained a considerable quantity of charcoal; for it is very black, and it soils the fingers. In the midst of this sort of basalt, I observed masses of coarse granular felspar, with some grains of quartz intermixed; so that at first I took them for fragments of granite imbedded in the basalt. But these felspar masses are also vesicular in structure, especially next to the basalt; a fact which satisfied me of their contemporaneous formation. They are often more than a decimetre in length; and although they abound in vesicles, are composed of granular distinct concretions, and are likewise intermixed with grains of quartz, they still possess nearly the form of those crystals of felspar that are found in porphyries; the only difference being, that here the angles and edges are rounded. The same regularity of shape is observable in similar masses of felspar contained in the basalt of the Riesengebürge.

---

* The little work here recommended by M. Daubuisson, and before alluded to, p. 130, has been excellently translated by Dr Charles Anderson of Leith. The translation was published by Messrs Constable & Co. in 1810.   T.

It is scarcely necessary to add, that when I express some surprise at the regular form of these masses, I am far from regarding them as crystals.

Note *f.*—(p. 89.)

Passage of Clay into Wacke and Basalt in Scotland.

In Scotland, the transition from Clay into Wacke and Basalt, as described in the text, may, in some places, be distinctly traced; and we frequently observe Trap-tuff passing into Wacke, the Wacke into Amygdaloid, and this latter into perfect Basalt. These transitions were first noticed in this country by Professor Jameson; and to him I am indebted for having particularly called my attention to two or three striking examples, which I shall here specify; all of them being taken from a small portion of the coast of Fife, opposite to Edinburgh.

Dunearn Hill, near Aberdour, is composed of trap-tuff, wacke, and basalt. The Trap-tuff forms the base and middle part of the hill; and the careful observer will, as he ascends, perceive this rock gradually refining into Wacke; and this again, passing into Basalt, which forms the upper part of the hill.

The small hills situated immediately behind the Bin, a hill at Burntisland, are composed of trap-tuff, wacke, amygdaloid, and basalt. The lowest part of these hills is composed of Trap-tuff: this tuff gradually passes into Wacke; the wacke into Amygdaloid; and this latter into Basalt.

On the coast, stretching from King's-wood-end, a little to the east of Burntisland, to Kinghorn, similar transitions may be traced.

In summer 1812, I had an opportunity of accompanying Professor Jameson and his pupils in an examination of the rocks along the shore between Kinghorn and Kirkcaldy. We found them to be flœtz-trap rocks, —amygdaloid, basalt, trap-tuff, wacke, limestone, and slate-clay. In several places, the Professor pointed out to us the Trap-tuff becoming gradually smaller granular, and passing into Wacke; the wacke assuming the amygdaloidal character, and passing into Amygdaloid; and this latter rock gradually becoming more compact, darker coloured, the vesicular cavities disappearing, till the transition into Basalt was completed.

I cannot help here remarking, how pleasant it was, and how well it augured for the progress of mineralogical science in Scotland, to see upwards of forty students, many of them provided with mineralogical bags and hammers, and all of them zealous and attentive, thus accompanying their Teacher in the investigation of the mineral scenes presented by Nature herself.

The rocks are here so well exposed by the action of the sea, that in the short space of three miles, we saw no fewer than sixty-five alternations. These, I shall here enumerate, from notes taken at the time; beginning at Pettycur and proceeding eastwards; the general direction of the beds being S. S. W.; and the dip, north of east:

1. Basalt.
2. Limestone.
3. Amygdaloid, with a base of wacke,—the general base of the amygdaloid here.
4. Limestone.
5. Flinty-slate.
6. Slate-clay.
7. Bed of sand and clay.
8. Bituminous shale.
9. Columnar basalt, containing fibrous zeolite; close by Pettycur Inn.
10. Amygdaloid.
11. Slaty rock, thick bed, consisting of bituminous shale and tuff.
12. Basalt.
13. Slaty rock, like No. 11.
14. Basalt.
15. Amygdaloid.
16. Basalt.
17. Amygdaloid.
18. Basalt, finely columnar; on the shore halfway between Pettycur and Kinghorn.
19. Amygdaloid.
20. Bed of slate-clay, bituminous shale and tuff.
21. A very thick bed, composed of basalt mixed with amygdaloid, both waving; the basalt generally tabular. In one place, the basalt contains iron-sand, and forms the *leck* sometimes used for oven-stones: in another place tuff with a base of wacke, occurs in the basalt. The amygdaloid is much traversed by minute veins of calcareous spar.

22. Slaty rock, like Nos. 11. & 13.
23. Slate-clay.
24. Sandstone, with numerous vegetable impressions; immediately at the town of Kinghorn, and terminating below the churchyard.
25. Slate-clay.
26. Amygdaloid.
27. Thin bed of sandstone.
28. Amygdaloid.
29. Quartzy sandstone; next the sea, four feet thick; diminishing in thickness inland, to two or three inches, and there passing into quartz.
30. Amygdaloid.
31. Quartzy sandstone; thin bed.
32. Amygdaloid.
33. Quartzy sandstone, along with slate-clay.
34. Amygdaloid, with tuff.
35. Argillaceous sandstone.
36. Bed composed of amygdaloid and basalt, with portions of limestone and of tuff.
37. Slate-clay, very much traversed by minute veins of calcareous spar.
38. Congeries of beds, consisting of three beds of Limestone and three thin beds of Slate-clay.
39. Amygdaloid, containing a remarkably large drusy cavity, full of calcareous spar; and a thin bed of a reddish substance, approaching to wacke, subordinate to the general

mass.—(About half a mile inland, near the high road, a quarry has been opened through this bed of amygdaloid, to get at the limestone below, *i. e.* No. 38.)

40. Slate-clay, full of petrifactions, chiefly a-nomiæ.

41. Bituminous shale.

42. Slate-clay, a red variety.

43. Limestone, formerly worked ; the kilns still remaining.

44. Amygdaloid ; the limestone thus resting on amygdaloid, and being covered by it.—Apparently subordinate to this bed of amygdaloid, is a small bed or tabular mass of limestone containing flint.

45. Limestone.

46. Slaty sandstone, containing numerous specks of mica, and coaly particles.

47. Amygdaloidal greenstone.—(The coast here, taking a turn, we no longer walk along the line or stretch of the outgoings of the beds, but cross them at right angles, so that their thickness is now much more easily ascertained.)

48. Sandstone, very clayey and slaty.

49. Greenstone, like No. 47.

50. Sandstone, with vegetable casts ; stems and bark of some extinct or unknown palm. The sandstone composing this bed, is disposed in thin partial layers, some of which are horizontal, others inclined.   A thin

bed of slate-clay occurs in the sandstone.

51. Limestone, full of petrifactions, chiefly en-
trochi; similar to what is called *Fife
marble*.

52. Sandstone, in which the appearance noticed
at No. 50., is still more distinctly seen; the
structure of the stone being often much inclined or waved, while the whole forms a
nearly horizontal bed.

53. Slate-clay and clay-ironstone.

54. Bituminous shale.

55. Limestone.

56. Slate-clay.

57. Sandstone.

58. Slate-clay.

59. Sandstone.

60. Slate-clay.

61. Greenstone.   In this greenstone, hollows occur, which are occupied by the sandstone;
and in one place the greenstone is seen covered by saddle-shaped layers of the sandstone and slate-clay.

62. Sandstone.

63. Thin bed of slate-coal.

64. Limestone, roof of the coal.

65. Greenstone.

T.

NOTE *g.*—(p. 94.)

Passage of Basalt into Greenstone in Scotland.—Basaltic Porphyry.

GREENSTONE is a very frequent rock in Scotland, particularly in the middle district of the country, that is, in the tract contained between the Grampians and the great southern range of mountains which extends from St Abb's Head, by New Galloway, to the coast of the Irish Sea. It is, as is well known, composed of hornblende and felspar; but Professor Jameson is of opinion, that augite frequently takes the place of hornblende in its composition. When the felspar predominates, what the Professor has termed *Sienitic Greenstone* is produced: When the concretions are small, and the hornblende predominates, the rock is named *Basaltic Greenstone:* and when nearly the whole mass is composed of imperfectly crystallized hornblende, it forms *Basalt.*

Salisbury Craigs, near Edinburgh, afford good examples of common and of sienitic greenstone. The entire mass of rock, which forms the bold semicircular cliff, seen from the city, consists of a thick bed of common greenstone, resting on and covered by beds of quartzy sandstone: and, in the greenstone, narrow veins of sienitic greenstone, some hundred feet in length, may in several places be observed. Basaltic greenstone occurs in the county of Fife; the small wooded hills in the park belonging to the Earl of Moray's seat of Donibristle, being chiefly composed of that rock.

The gradual transition from basalt into greenstone, is observable in two small isolated hills in East Lothian, called North Berwick Law, and Traprain or Dumpender Law. Dr Ogilby, who carefully examined that district of country, found that the lowest or oldest rock was uniformly a coarse conglomerate; that this graduated into amygdaloid; which, in its turn, either passed at once into basalt, or, in some instances, first through claystone; and this either into basalt or compact felspar: these, again, passed into porphyry-slate, and the passage into greenstone at the top of the above-mentioned hills, was indicated by interposed particles of hornblende.

Not only does flœtz-greenstone appear to pass into basalt; but *Sienite,* a member of the Primitive and Transition series, seems in some instances to pass into a rock which Professor Jameson has proposed to call *Basaltic Porphyry.*

Ben-Nevis, the highest mountain in Great Britain, which has been well described by Dr Macknight, appears to contain basaltic porphyry:

" It is at this spot," (where the guide asks travellers to rest and refresh themselves, before attempting the most difficult part of the ascent,) " that the external character of the formation, appears to undergo a remarkable change. On the opposite, or south side of the rivulet, two fronts of the rock *in situ,* projecting from the debris by which they are surrounded, present themselves, the one at the distance of some hundred feet

above the other, in the line of ascent. Of these, the
inferior is a small granular sienitic-porphyry, with
scales of mica, and a reddish colour; but the higher
rock is unexpectedly found, on applying the hammer,
to consist of a greyish-black substance, sometimes inclining to a deep green, with an uniform texture, which
seems at first sight in its characters and fracture to resemble basalt.

" Struck with this unlooked for appearance, and
comparing it with what I afterwards found at the summit, I conceived at first, and for some time, that the
observation thus made was the discovery of a flœtz-for-
mation, consisting of clinkstone over porphyry-slate
and basalt, and resting immediately on sienite. I therefore applied myself with eagerness to find out what
might be considered as the junction or line where the
transition takes place. But the intermediate space, the
bed of the rivulet, and the surrounding acclivity in
every direction, are so completely covered with fragments, that the line sought cannot be discovered here-
abouts. So that, after spending a long time, and returning on purpose another day to prosecute the search,
I was obliged to abandon it without success.

" At the same time, although on this face of the
mountain, the quantity of debris from the higher rocks
is prodigious, and descends probably a little below the
upper part of the sienitic mass, the experienced eye can
easily trace the horizontal line, where the fragments of
the subjacent felspar-porphyry begin to mix with those
of the overlying dark-coloured substance. It runs
along the acclivity towards the north shoulder, and is
evidently but a small distance under the presumed junc-

tion. I may further observe, that all the rocks at this part of the ascent, are so little uncovered, and decompose in such a manner, that it is difficult to ascertain whether or no they are disposed in beds, and if they are so, to determine precisely their dip and direction. Splitting, indeed, in rhomboidal masses, they appear in one view to consist of layers or strata, which maintain a pretty regular bearing, east of north, with an inclination of 75°; dip north of west. But another view presents uniform lines of separation in the mass, which suggest the idea of a dip and direction entirely different; a circumstance, which is probably owing to the structure of the stone, as occurring in tabular distinct concretions.

" It then struck me, that the junction I had sought in vain, might perhaps be found laid open on the opposite side of the mountain, along the front of the precipice * "

Afterwards, p. 342, the Doctor proceeds, " In this lonely region," (the summit of Ben-Nevis,) " my surprise and delight were raised to the utmost, by discovering the line of transition or junction between the different coloured masses, laid bare on the front of the precipice near its foot, and stretching horizontally for almost a mile. The beginning of the line, at the east projection is accessible, by climbing up the broken rocks; and, on advancing a little farther in this direction of the hollow, we find it open to a splendid view

---

* Memoirs of the Wernerian Natural History Society, vol. i. p. 330.

of the same beautiful appearance, which, along the face
of the perpendicular section, exhibits the structure of
the whole summit. This line of apparent junction,
though horizontal, is not straight, but a minute angular
zigzag ; and when closely inspected, is found to be, not
the separation of two different rocks, but merely the
passage of the same rock into a different colour. For,
the gradual transition of the one substance into the
other, with no change on any of the characters, excepting the colour, demonstrates the identity and continuity
of the formation. We see it distinctly at the spot under the north-east front of the precipice, where the
transition is first observed."

" As the upper part of this formation has certainly
an unusual and unexpected appearance, in a geognostic
point of view, I shall now enter a little more particularly into the consideration of its oryctognostic characters, for the purpose of determining its nature and relations.

" This rock, as formerly observed, contains in general crystals of felspar. It occurs, however, in different places, without the porphyritic structure : and has
then a strong resemblance in some instances to basalt,
but more frequently to clinkstone ; as at the top, where,
if struck with a hammer, many of the fragments or
loose pieces, ring like metal. Its fracture exhibits the
varieties of the splintery, the flat conchoidal, and the
foliated ; and its colours, those of the greyish-black,
and the dark greenish-grey. But it does not appear to
be ever vesicular, or translucent on the edges ; it contains no traces of olivine or augite, so far as my observation extended : nor is the principal fracture slaty in

P

the great, a character without which the porphyritic
varieties cannot be considered as porphyry-slate.

   " True clinkstone and basalt, indeed, are substances
which belong to a newer and very different æra of formation, and do not seem to be so purely chemical in
their nature as the rock we are considering. But the
affinity of external characters now pointed out, illus-
trates one of the principles of the Geognosy *, and appears strongly to support a conjecture of my friend
Professor Jameson, that clinkstone and felspar are nearly allied : because the geognostic position and relations,
along with the splintery fracture, of the Ben-Nevis-
rock, demonstrate, that it has a strong affinity to compact felspar tinged with hornblende ; and, from a variety of observations made in Arran, Dumfries-shire,
and the vicinity of Edinburgh, Mr Jameson has inferred, that clinkstone-porphyry passes into compact-fel-
spar and claystone. The great distinction betwixt the
substances of this genus, which belong to the primitive
and flœtz periods, seems to be, that the former in general are more crystalline ; something a-kin to the relation which subsists betwixt primitive and transition
or flœtz-limestone."

   Von Buch, a well known and accurate observer, describes a transition from porphyry into basaltic porphyry,

----

   * " The resemblance of the newer porphyry to the newest
flœtz trap formation, is deserving of attention. The points of
the agreement are, *in the stone itself,* in the structure of the
rock," &c. Jameson's *Geognosy,* p. 138.

which he remarked near Christiania in Norway * ; and Haussman, likewise an excellent observer, describes basaltic porphyry among the porphyry and sienite rocks of Norway †.

'T'

Note H.—(p. 97.)

Constituent parts of Basalt.

IF the results of the analyses which have been published of hornblende and felspar, and of basalt, had uniformly agreed, it might have been possible to combine the two former in such proportions as that the alloy produced should have been of the same nature as basalt : unluckily, we do not yet possess data sufficient to establish such an equation.   Assuming, however, the analysis of felspar by Vauquelin, and that of hornblende by Hermann ; by adding together three-fourths of the constituent parts of hornblende, and one-fourth of those of felspar, we get a result nearly approach-

P 2

* *Reise durch Norwegen und Lappland*, B. 2. ;—a valuable work, of which a translation into English is soon to appear.

† *Reise durch Scandinavien, in den Yahren* 1806 *und* 1807, B. 1. 308.

ing to the analysis of basalt by Klaproth; as may be seen in the following table, in which the analysis of basalt by Kennedy is also set down :

TABLE.

	VAUQUELIN. Felspar.	HERMANN. Hornblende.	Alloy of 4th felspar, and ⅔ths hornblende.	KLAPROTH. Basalt.	KENNEDY. Basalt.	
Silica, . : . : .	62.83	37	43.46	44.50	48	
Alumina, . . .	17.02	27	24.50	16.75	16	
Oxide of iron, .	1.00	25	19.00	20.00	16	
Lime, ...	3.00	5	4.50	9.50	9	Volatile matter.
Magnesia, . . .	0.00	3	2.25	2.25	0	
Alkali, ...	13.00	0	3.25	2.60	4	
Water, ...	0.00	0	0.00	2.00	5	
Muriatic acid, .	0.00	0	0.00	0.01	1	
Oxide of manganese, ...	0.00	0	0.00	0.12	0	
Carbon, ...	0.00	a trace	a trace	a trace	3	
Loss, ...	3.15	3	3.04	2.27	1	
	100	100	100	100	100	
	STRUVE,	HAÜY,				
Specific gravity,	2.55	3.25	3.076	3.066		

It will be observed, that I regard this table as a mere sketch, or an approximation to the truth; and am aware that several objections occur to the synthetic proportions here set down; for instance the felspar analysed by Vauquelin, was the green felspar of Siberia, and not that found in greenstone, which seems to differ somewhat from common felspars, and is more fusible : further, the alkali found by Vauquelin in fel-

spar, was potass, while it was soda that Klaproth ex-
tracted from basalt; it is to be considered, however,
that some chemists have supposed that these two al-
kaline substances may be merely modifications of the
same base,

~~~~~~~~~~~~~~

NOTE I.—(p. 99.)

Answers to some Objections.

IT may perhaps be argued, that if the basalt, which
now appears on the summits of the mountains of Saxony, be really the remains of a sediment or deposition, from a solution which covered the country; that
sediment must necessarily have covered places at a lower level, and basalt should be found at all such places.

I might reply, in the way of analogy, that gneiss,
mica-slate, porphyry, and sandstone, exist on the most
elevated points of the globe; and that consequently the
whole earth has been covered by the solutions which
produced them; yet vast countries may be travelled
over, without the least vestige of these mineral substances appearing. I shall specify one very striking
example. M. Ramond found a limestone rock full of
very entire shells, situated on the ridge of the Pyrenees,
between nine and ten thousand feet above the present
level of the sea. There cannot certainly be the least
doubt that this limestone has been deposited from an

aqueous solution in a state of tranquillity, or in other words, from an ocean, which must necessarily have covered the whole earth, with the exception of a few lofty summits: and yet there are vast tracts of country, where not the slightest trace of a calcareous deposite is to be seen.

I shall now make a direct answer to the objection. I may begin by mentioning, that I have myself seen basalt, on the summit of the Giant Mountains, at the height of more than 4000 feet above the level of the sea: it is evident, therefore, that the solution which produced basalt, must have covered nearly the whole globe, or must have been *universal:* but it does not follow, that the precipitate or deposition, must also have been *universal;* for,

1. The solution may very well be supposed not to have contained, at the same time, and throughout its whole extent, the same constituent parts, or not to have contained them every where in the same proportions. Indeed, it is known, that at the present day, the sea, in different latitudes, does not hold the same substances in solution, at least in the same quantity.

2. Local causes, such as a subtraction of caloric, or the presence of certain precipitants, may have affected only particular parts of the solution, and produced a deposition in one place, while there was none in another.

3. These causes, acting differently in different situations, may have favoured the union of particular constituent parts in a certain proportion, while, at a distance, there may have been formed, from the same elements, a

combination of a nature considerably different. For instance, it is quite possible, that from the same general solution which deposited basalt in Saxony, may have been precipitated in America the extensive tracts of porphyry-slate observed by M. Von Humboldt *, and which, according to the analyses of Klaproth and the geognostic observations of Werner, is nearly related to basalt, (§ 47. 62.)

4. In the vast sea, or general solution alluded to, there must necessarily have been currents and agitated parts, while other portions remained in a state of calm more or less perfect. Hence another evident cause of difference in the nature of the precipitate. In a calm situation, greenstone may have been forming, at the same time that basalt was precipitated from a place subject to agitation. In the same way, in the production of calcareous rocks, a granular limestone may have resulted, where the solution was sufficiently tranquil to permit crystallization; while compact limestone was produced from the agitated waters. It has been already remarked, (§ 30,) that greenstone bears the same relation to basalt, which granular limestone bears to limestone that is compact.

5. The currents and motions alluded to, may have obstructed precipitation in one place, and favoured it in another; they may, in some countries, have washed away the various sediments, shortly after their deposi-

P 4

---

* *Journal de Physique*, tom. 53. p. 56.

tion, and may have heaped them together in other places.

It appears therefore most probable, that the basaltic precipitate never, at any one time, entirely covered the globe: we have seen that any particular part might have been removed, almost the moment after its deposition. But it must have been after the waters had retired, that the great basaltic bed was shattered, divided, and so much destroyed, as we now see it.

This disintegration and destruction of rocks, is a fact well established in geology. The grey-wacke, sandstone, puddingstone, and breccia, which frequently constitute entire countries, and form great chains of mountains; the vast deserts and plains of sand which occupy nearly a quarter of our globe; all the collections of sand and gravel which exist at the bottom of the sea; together with the beds of clay and of earth, which at present cover the surface of rocks; are evidently nothing but the remains of extensive mountains which formerly existed. To furnish so vast a quantity of debris, how immense must have been the detritus which they have undergone! Now, as the basaltic mass once covered almost all the other rocks, these last could not have been attacked till after the destruction of the basalt, or of its constituent parts; and, therefore, if the shattered and incomplete state of the basaltic deposition present any thing surprising, it is rather, perhaps, to find that certain portions have resisted destruction, than that the greater part has yielded to it.

What, then, it may be asked, have been the causes of this destruction? Without directly answering this question, it is sufficient for my purpose to be able to state, that so stands the fact.   To help to explain, however, the broken aspect of the basaltic deposition, I may remark, that it is very possible that the solution itself which produced it, might destroy it in a considerable degree during its retreat, especially if this was rapid and sudden.   The observations of Saussure, and of some other geologists, prove the violent effects produced by a flood which formerly passed over the earth's surface; the waters have left imprinted on the rocks the traces of their violent passage; the Genevese naturalist could not mistake them in the Pays de Vaud.   But, after all, the grand cause of the disintegration of rocks, is the destructive action of the elements: however insignificant such a cause may appear at first sight, by acting continually and without interruption during a long series of ages, it cannot fail to produce mighty effects.   I may here repeat what has been elsewhere observed on the same subject,—That Nature has Time at its disposal, and that a finite effect repeated an infinity of times, is in reality an effect infinitely great.

If the objections be pushed still further, and if, while I allege that basalt is a precipitate proceeding from a solution which formerly covered the places where it is now found, I be required to specify the nature of the solvent, to tell whence it came, and from what quarter it procured the constituent parts of basalt; to assign a cause for the precipitation; to say what has become of the menstruum, and where it now exists;— to all such queries, I answer, that I do not interfere with remote causes, but confine myself to facts, and to consequences which immediately flow from them.   I

assign to basalt, an origin similar to that of other
rocks ; and it is evident, that the above objections are
equally applicable in the case of all rocks.   The geognost must of necessity content himself with establishing
facts ; and, therefore, what I shall answer farther, will
be very short.—The spheroidal form of the earth, appears
to those best versed in astronomy and physics, a proof
of its original fluidity.   Crystallisation necessarily supposes a previous solution ; and therefore the crystalline appearance of granite, gneiss, and other rocks, as
as well as the numerous crystals which they contain,
shew that the constituent parts of those mineral masses
which form our globe, have at one time been suspended in a menstruum.   The stratified structure of many
mountains and rocks, indicates that they have been
formed by a succession of precipitates or sediments.
The existence of beds of sandstone and puddingstone
at great heights, and the shape and arrangement of the
rolled masses included in them, prove beyond controversy, that currents and waves once covered those heights.
Lastly, The petrifactions which abound in many rocks,
and particularly the remains of marine animals, some
species of which still exist in the ocean, entitle us to
believe that, at least at certain epochs, the vast solutions from which the substance of the rocks has been
precipitated, must have been analogous to our present
seas.   These are facts ; and our inability to explain
their cause, certainly does not invalidate them.

BASALT OF THE SCHNEEGRUBE.   285

Note K.—(p. 76.)

Basalt of the Schneegrube.

On the summit of the Riesengebirge, or Giant Mountains, about 4000 feet above the level of the sea, there is a deep hollow fronting the north, and forming the commencement of a valley. The rock which constitutes the whole mountain, is a coarse-grained granite. In this place, it appears to be stratified, or rather divided into horizontal tables, which are a little inclined towards the north-east. The hollow is known by the name of the Schneegrube, or Snow-pit. It really does contain snow the whole year round; and the streamlet which arises from the gradual melting of this, has probably been the principal means of forming the valley below. The pit may be about 150 feet deep; and it seems to have been produced by a sinking of the strata. In the western side of this pit, a mass of basalt appears, of an elongated form, and placed with its back to the granite. Its two sides are separated from the granite by fissures. It is about twenty-five feet in breadth; and its length is nearly equal to the depth of the pit. It is impossible to see whether it runs farther down. It is laid against the granite; and in one place the surface of contact is distinctly seen. The thickness of the mass from that contact-surface to the exterior face, is from ten to twelve feet. Possibly it may formerly have been somewhat thicker ; but it has never been large ; for no vestige of it can be traced on the eastern side of the pit. If the view thus to be got of this basaltic mass were sufficient to entitle us to draw any conclusion concerning its shape, (which I scarcely think it does,) I would say that it seems to form a sort of parallelipiped, more than a hundred and fify feet in length, five-and-twenty in breadth, and twelve in thickness. What sort of repository *(gîte)* may this mass be considered as forming ? It cannot be ranked as a bed ; for it is vertical in position, while the stratification of the granite, (if stratification be here admitted,) is nearly horizontal. It cannot be classed as a vein ; for its length, in the line of its direction, can never have exceeded 150 feet, and its thickness is from nine to twelve feet : While these are evidently not the dimensions of a vein, it is equally destitute of the form and structure of a vein. Perhaps this mass may constitute a repository similar to the *Butzenwacke* at Joachimsthal, described by Werner in his Theory of Veins *. The basalt of this mass is very compact, but it contains some vesicular cavities, partly empty, but more generally filled with fibrous zeolite. The basalt is likewise traversed by small veins of zeolite, only about a line in width, but several inches long : these have probably been fissures, afterwards filled up. The cavities, further, contain green steatite and calcareous spar. Some grains

* Sect. 128. Page 233. of Dr Anderson's translation.— Professor Jameson, who has seen the basalt of the Schneegrube, considers it as an upfilling, *(ausfüllung.)* T.

of olivine also occur in this basalt.   Masses of felspar
are common, both in separate grains, and intermixed
with particles of quartz : these masses incline to a
regular form, in the same way as those already mentioned, (Note E.)   The mass of basalt is divided at the
base into irregular prisms, and, near the top, into
tables, which form groups ; the laminæ in each group
having a different inclination.

~~~~~~~~~~~~~

Note L.—(p. 99.)

Opinion of Werner concerning the Formation of Basalt and of
   Trap-Rocks in general.

It may not be unacceptable to the reader to have a
short account of Werner's ideas concerning the formation of basalt and other trap-rocks, suggested to him
by a long train of observations, and profound reflections.

He remarks,
1. That basalt is often accompanied by wacke, clay,
&c. (Note D.) ; that in some places it appears along
with greenstone, and in others with porphyry-slate ,
and that these different substances always occur together.   He thence concludes, that they have been
formed at the same epoch, or, in his geognostic language, that they are *of the same formation.*

2. A gradual passage of the trap-rocks into each other is observable, clearly evincing their natural alliance. The shades of change, by which they pass successively into each other, he considers as proving that the sediment or precipitate from which they were produced, had by degrees altered its nature; in a word, that they are all the product of the same solution, and belong to one family of rocks. To the rocks of this family or formation, he gives the general name of Flœtz Traps.

3. In this formation, the substances which were first formed or deposited, and which now occupy the lowest station, are the coarsest, such as gravel and sand : the particles of those which were subsequently deposited, become gradually finer, and cohere more strongly to‧gether; these are clay, wacke, and basalt; those in which the combination between the particles is most intimate, as porphyry-slate, and those others in which it becomes a true crystallization, as in greenstone, are generally in the upper part of the series, having been last formed. From the order in which the substances have been deposited, he concludes, that at first the solution had been much agitated, and that only the coarsest particles could then be deposited ; but that it had gradually become more and more tranquil and pure ; and had at length attained that state of calm, favourable to crystallization.

4. This formation of trap-rocks had been preceded by a revolution in our globe, which had greatly destroyed or disintegrated the rocks previously existing, producing gravel, sand, and other debris.

5. Among the lowest or earliest products of this formation, are frequently found beds of bituminous wood, or of coal: it thence appears, that anterior to its date, the surface of the globe must have been clothed with forests.

6. As trap-rocks exist at great heights, the waters of the solution must have risen very high. It does not, however, appear that they had covered the highest mountains; for no basalt or other flœtz trap-rocks are found on the summits of these.

7. As the flœtz-trap formation covers all other rocks, even the newest, (collections of alluvial matter excepted,) it must have been the latest formation; and the revolution in nature which preceded it, is the last of which any vestige is imprinted on the mineral surface of our planet.

From these observations, Professor Werner concludes, that the globe of the Earth is of remote antiquity; that its surface was inhabited by animals, and covered with vast forests, when it underwent a great revolution, perhaps the last of several which it has experienced; that this revolution occasioned the disintegration of many of the rocky masses already existing,—the total destrcution of the forests,—and was followed or accompanied by a mighty inundation, which rose to a height, equal perhaps to that of the highest mountains; that this immense and necessarily raging sea produced accumulations of gravel and sand, over which, when it had somewhat abated of its agitation, were deposited the earthy, clayey, and bituminous particles with which it was charged: that as the water became more and more tranquil and pure, the precipitates had become less earthy, and the union between their particles more intimate; wacke, basalt, greenstone, and porphyry-slate, being successively produced, as it approached to that state of calm and purity, favourable to crystallization.

The remarkable reserve of Werner, has prevented him hitherto from giving to the world any full exposition of his geological doctrines; but what I have now stated, I know to accord with his sentiments. I cannot help here remarking, that the long-continued practice of observing Nature, and of observing with the most scrupulous exactness, seems in some measure to have let him into the secret of her operations; insomuch that the hints which he throws out as mere conjectures in geology, appear to me much more deserving of attention, and much more consonant to what we know with certainty, than the laboured hypotheses of most other geologists.

It may not be amiss here to explain what is meant by a *suite of formations*.

It will be recollected, that all rocks which have been formed at the same epoch, are considered as belonging to one formation. Thus, the coal-formation contains not only coal, properly so called, but certain kinds of sandstone and slate-clay, which almost always alternate with the beds of coal, and which have consequently been formed at the same time. But the formation of the same mineral substance, or of a substance nearly identical in chemical composition, has been repeated at different epochs. For example, we find beds of lime-stone in gneiss, in grey-wacke, in sandstone, &c.: now, all of these limestones belong to different formations, and form what is termed the *limestone formation-suite*, which thus includes, the primitive, transition, flœtz, and shelly limestones, with chalk, and calc-tuff found in alluvial soil. The terms applied to the different members of such a formation-suite, indicate in some measure the period, or at least the order, in which they have been formed. The primitive limestones are large-grained, of a crystalline aspect, and translucent : In those which occur along with slate-clay, the grain becomes smaller, and the translucency diminishes : almost all the flœtz limestones are entirely compact, the fracture is slaty, and they are only translucent on the edges : in proportion as they are newer they become more earthy and more opake, till they appear completely so in the chalk formation. What a contrast between the extreme limits of this limestone-suite,—the fine marble of Carrara and chalk or calc-tuff ! The first formations of limestone have been pure and crystalline ; they appear to have gradually become turbid, and mixed with foreign substances ; till they at length degenerated into mere sediments or mechanical deposites.

We find in the trap rocks a series analogous to that which we have just noticed. We have seen that basalt passes into greenstone, a rock composed of grains of basaltic hornblende and felspar. A similar greenstone exists among the primitive rocks, and it there passes into another rock, entirely composed of granular hornblende. This is the first link of the chain of rocks having a basis of hornblende. From this point, till we arrive at basalt, or at wacke, we have an uninterrupted gradation. All the rocks of this suite are denominated

traps by Werner ; and according as they are found associated with primitive, transition, or flœtz rocks,
they are called *primitive, transition*, or *flœtz traps*.
The remark just made concerning the limestone-
suite, is likewise applicable here : the oldest trap-
rocks exhibit a completely crystalline structure : the
kind of structure gradually changes, and degenerates,
until it becomes earthy, as may be seen in basalt, and
still more distinctly in wacke.

Note M.—(p 106)

Impossibility of the Volcanic Origin of the Basaltic Rocks of
Saxony.

It would doubtless be no easy task to fix on a spot
where any single crater might have existed, which
could possibly cover all the mountains of Saxony
with its lavas. These mountains, it has already been
remarked, are joined to those of Lusatia ; and these,
again, to those of Silesia ; all of them forming but one
great chain : the beds or masses of basalt which occur on
them, are every where of the same character, and bear
the same relations to the other rocks : they are evidently the remains of one great bed, or, in the volcanic
view, they must have been produced by one immense
stream of lava. It may be assumed, that the crater
must have been situated on a level at least with the

highest portion of the basalt or lava. This highest portion is at the Snow-pit on the summit of the Giants Mountain, (already mentioned, p. 235): but the crater could not be there; for all the basalt or lava to be seen at that spot, is a mere patch, a few feet in thickness and length, and completely surrounded by granite. The crater, therefore, ought to be found at a still higher point of elevation. Yet it will not fail to be remarked, that the Snow-pit is situated on the ridge which forms the most elevated line between the Baltic Sea and the plains of the Danube, more than 4200 feet above the level of the sea, and about 4000 feet above the neighbouring valleys; that there are very few spots of the ridge more elevated, excepting one eminence composed of granite, called the Giant's Head (Riesenkoppe), which rises about 600 feet higher; and that the Giants' Mountains in reality constitute only one great mountain mass, above thirty miles long, but very narrow. If, therefore, the basalts of Lusatia, Silesia, and Saxony, were produced by a single volcano, the lava must have issued from the most elevated point in Germany, and must in that view be held to have opened a passage to itself precisely at the point which offered the greatest resistance! It is to be presumed, that the volcanic fire was under, not within the mountain; and in this case, it seems strange that the lava did not flow from the base, directly into the plain, instead of forcing its way through granite, to the summit.

It may perhaps be alleged, that great changes have taken place, and that, what is now an elevated summit, may, before the wearing away of the surrounding rocks, have been comparatively low. But this I cannot admit.

The natural effect of the causes which are continually operating on the crust of the globe, is to raise the level of the low land, by lessening that of the more elevated parts; so that in former times the difference of level between the base and summit of the Giants Mountain, must have been still greater than now, and my objection must then have been still more insurmountable. I conclude, therefore, that it is entirely impossible for us to suppose, that a crater could ever have existed, from which all the basaltic masses of Saxony could flow.

Note N.—(p. 113.)

Comparative fusibility of Basalt, Greenstone, and Lava.

Sir James Hall observes, that " *whinstone* (that is, greenstone) is composed of black hornblende, and of a substance resembling felspar, but much more fusible: this substance melts more readily than the hornblende with which it is mixed."

The experiments of Sir James Hall on the comparative fusibility of basalt, greenstone, lava, and some other substances, appear to have been made with great care, a pyrometrical piece being kept in the crucible, close by the substance melted. The results which he obtained, therefore, are much more to be depended on than the loose calculations of Saussure, who does not

seem to have attended to the fact, that the several varieties of felspar and hornblende are fusible in very different degrees. Sir James found, that well characterized greenstone began to melt at a temperature equal
to 40.55 degrees of Wedgwood's pyrometer; a fragment of the basaltic columns of Staffa, at 38°; lava
from Etna, having a stony aspect, at 32°; and lava of
Vesuvius, eruption 1785, with a vitreous aspect, at 18°.
According to Guiton, copper melts at 27°, and cast
iron at 130°, of Wedgwood's pyrometer.

Note O.—(p. 120.)

Difference between Basalt and Lava.

I am by no means satisfied concerning the perfect
resemblance which has been alleged to exist between basalt and compact lava; all specimens taken from undoubted lava-streams, having more or less of a vitreous aspect. Even the excellent experiments of Sir James Hall,
are not conclusive on this subject. Professor Jameson
informs me, that he found no difficulty in distinguishing
the specimens of melted basalt from real lavas. I understand that the melted basalt, slowly cooled, produces a stony substance of a deeper colour than natural basalt; and also much harder, giving sparks with
steel, which natural basalt does not. At Edinburgh
the resemblance is said to have been complete; but at

Geneva, there appeared to be sensible differences. Some mistakes may easily have arisen from mistaking for real lava, a fragment of basalt found on Vesuvius, which perhaps had been torn from the walls of the subterraneous caverns and ejected. The similarity of colour might readily deceive any one but a practised mineralogist.

One passage of Sir James Hall's paper, seems to shew, that such mistakes had sometimes happened: " It is generally supposed," he says, " that some lavas of Ætna contain calcareous spar and zeolite; but this I conceive to be a mistake. It is true, as I have seen, that many rocks of Ætna contain these substances in abundance: but in my opinion these rocks are no lavas, but are the same with our *whins*, (greenstone, basalt, wacke, &c.) in every respect. One district of Ætna is decidedly of this description; and vestiges of the same kind occur in other parts of the mountain. In one place fossil-coal has been found, and in another we saw marine shells *."

Sir James Hall visited Ætna, in 1785, in company with Dolomieu. He appears, from his writings, to be a disciple of Dr Hutton, who supposes that granite, porphyry, basalt, and similar rocks, have once been in a state of fusion; and that, while thus in fusion, and for a long time after their congelation, they remained far below the then surface of the earth : they are therefore considered by Dr Hutton and his followers as a kind of subterranean or unerupted lavas.

---

* Transactions of the Royal Society of Edinburgh, vol. v. part i. p. 69.

~~~~~~~~~~

Note P.—(p. 123.)

Crystals uniformly dispersed through Basalt.

It appears to me, that a difference in specific gravity must, at least to a certain extent, have had an influence in determining the position of crystals or other foreign substances involved in fluid lava. The lighter sorts would float at the surface of the melted mass while yet in the volcano; they would maintain the same situation when the lava burst forth; we should still find them at the surface, and certainly not equally mixed through the whole body of the lava. No doubt the very flowing of the lava might in some measure derange the order of position which such substances would take by the laws of hydrostatics : but I do not think that the effect would be considerable; for, notwithstanding the most violent agitation of a rapid river, pieces of wood thrown into the stream will be observed constantly at the surface. It may perhaps be said, that the viscidity of melted basalt is such, that substances accidentally displaced could not readily regain the situation fixed by their specific gravity : but it is evident that this viscidity would equally tend to prevent the derangement of substances which had once taken their proper places ; and it affords no explanation whatever of the uniform dispersion of the crystals or foreign substances through every part of the mass.

~~~~~~~~~~~~~

Note Q.—(p. 127.)

Felspar and Hornblende in Lava.

THE mineralogists who have described Etna, make mention of crystals of felspar being observable in almost all the lavas of that mountain. But it may be questioned, if these are truly felspar. Fusibility is one of the distinctive characters of that mineral: it is composed of silica and alumina, nearly in the proportion of 4 to 1; it sometimes also contains a portion of alkali; and such a mixture must necessarily be fusible. The hornblende which occurs in the same lavas, is readily melted; more easily than the felspar, according to these authors. If, therefore, these substances are fusible by themselves, without addition, before the blow-pipe or in our furnaces; much more may they be expected to be so, in the vast fire of a volcano, in the midst of a bath, as it were, of various earths and metallic oxides, acting as fluxes. The homogeneous texture of lavas appears to me to prove, that the substances which enter their composition have been completely dissolved and melted; and that if isolated crystals occur in the midst of lava, such crystals must have been formed during the process of cooling and congelation.

Dolomieu himself mentions his having put into a furnace pieces of lava, containing crystals of felspar and hornblende, (but it is not impossible that these may have been pieces of real neptunian basalt, that rock almost encircling the lower part of Etna); and he admits that the same heat which melted the lava, melted the contained crystals, and that the product was a homogeneous glass. Still, however, he holds that the felspar and hornblende existed in the rocks which furnished the substance of the lava; and that these rocks had only been softened by the influence of the volcanic fire, as otherwise the crystals would have been deranged and altered.

He thus expresses himself concerning the felspar: "The scales and crystals of felspar, are not foreign to the rock which contains them; they have not been accidentally enveloped, but have been produced in the spot where they exist; having been formed by the attraction between similar particles of matter, while the whole substance was in a state sufficiently soft to permit such actions: I here speak of the fluidity which the rock must have possessed during its first formation in the humid way, not of the softness or semifusion which may afterwards have been produced in it by volcanic heat *."

* List of the lavas of Etna, printed at the end of the *Memoire sur les Iles Ponces,* 1788, p. 202.

Dolomieu, it will be remembered, thought it necessary to consider volcanic fire as of a nature altogether peculiar; but I see no reason for adopting that
opinion. The homogeneousness of the paste of lavas,
shews that the heat is sufficiently intense to melt all
the constituent parts; and it seems difficult to explain
how certain detached crystals should have escaped the
influence of that heat, and remained perfect. When
experienced chemists investigate the subject, it will probably be found, that the *matter of heat* in a volcano does
not essentially differ, in its nature or effects, from common *caloric.*

The appearances described by Dolomieu, (supposing
them correctly observed,) are not only at variance with
some established chemical doctrines, but, what is of
more consequence, with the first principles of hydrostatics. " The fluidity of the lava," he says, " has not
altered either the external form, or the lamellar structure of the felspar; nor have the crystals changed their
local position in the mass * :" and he mentions a lava in
which the crystals of felspar were separated about two-
thirds of an inch from each other, and spread near the
bottom of the mass with a considerable degree of regularity. But perhaps it may be questioned whether these
crystals were really situated in a lava: how is it possible, indeed, to conceive that bodies specifically lighter
than the fused matter in which they were suspended,
should have retained their respective positions, without
running together in groups, or endeavouring to rise to

---

* *Ut suprà,* p. 251.

the surface, during all the violent agitations to which the fluid must necessarily have been subjected; for, it is to be considered, that the lava must have existed for some time in the volcanic caverns in a state of fusion, and must there have been kept in motion by the continual escape of elastic fluids; that it must afterwards have been raised from the subterranean abyss, to the summit of a mountain, perhaps 9000 feet high, by an energy far surpassing any projectile force with which we are acquainted; and that it must have flowed from the elevated crater with the rapidity of a torrent.

I admit, however, that it is impossible and absurd to argue against facts; and if the mass in which these crystals occur, be a real lava, I would be inclined to regard this as one of those mysterious appearances in nature, which we must believe without being able to explain.

NOTE r.—(p. 141.)

Basalt and Greenstone alternating with Coal in West Lothian.
—Notice of the late Mr Williams, the Mineralogist.—Basalt and Coal in Mull,—and in the Faroe Islands.

Basalt and greenstone with Coal in Linlithgowshire.—
In § 47., Mr Daubuisson argues against the volcanic origin of basalt, from the relative position which it ge-

nerally occupies in regard to other minerals, such as limestone, clay, or coal.   He brings forward examples of its alternation with such mineral beds, and coming in contact with them, without any alteration being produced.   At the Meisner Mountain in Hessia, it rests, in one place, immediately on a bed of highly bituminous coal *, which, however, exhibits no indication of having suffered any alteration from contact with the basalt.

I understand that it is the true *Basalt* of mineralogists which occurs at the Meisner next to the coal.

---

* STRUCTURE OF THE MEISNER MOUNTAIN.

Greenstone.

Basalt.

Columnar coal.

Conchoidal glance-coal.

Pitch-coal.

Brown coal, often verging on pitch-coal.

Brown coal, often including bituminous wood
and earth-coal.

Bituminous wood.

Limestone.

Sandstone.

In some of the other examples given, such as most of those in Scotland, the rock called Basalt in Mr Daubuisson's Memoir, we know to be the *Greenstone* of Werner. But that is a circumstance which evidently does not affect the argument. Not only has basalt, in many places, been traced passing into greenstone, so as to prove that both have had a similar origin ; but Plutonists and Neptunists are agreed as to that point, and dispute only whether the *origin* has been igneous or aqueous.

The remark now made, is applicable to the instances quoted by Mr Daubuisson from Williams's " Mineral Kingdom," of " thick beds of basalt occurring between beds of coal," at Borrowstounness and in the Bathgate Hills, both in the county of Linlithgow . Mr Williams calls these thick beds " basaltine rock," which he explains to be the rock " commonly called *whinstone*," —a Scottish term which universally includes greenstone, and is most frequently applied to that rock. His words are : " Strata of basaltine rock are very common in many coal-fields in Scotland. There are several thick beds of this stone betwixt the different seams of coal at Borrowstounness, and one of them is the immediate roof of a seam of coal in that ground.—In the Bathgate Hills, south of Linlithgow, there are several strata of coal, and several strata of basaltes blended together, *stratum super stratum* "

I have been told that whinstone is not the immediate roof of any of the seams of coal at present worked at

* *Antea*, p. 140.

Borrowstounness. But Mr Williams may very possibly have observed it forming the immediate roof of some thin and insignificant seam of coal, which, though an instance equally conclusive in favour of the argument, might very readily escape the notice of those who interest themselves only about workable seams. It may further be remarked, that large beds of greenstone occur in the coal-field of Borrowstounness. A bed of this kind several yards in thickness, resting on sandstone and covered with the same sort of rock, is very well exposed on the shore of the Frith of Forth, about a quarter of a mile westward from the town.

From my friend the Rev. Mr Fleming of Flisk, who is well acquainted with the mineralogy of West Lothian, I learn, that the mines in the Bathgate Hills, where in all probability Mr Williams made his observations about forty years ago, have long ceased to be worked, and the appearances which he describes, cannot therefore, at this day, be distinctly observed. Many examples of the occurrence of beds of greenstone in the independent coal formation of Linlithgowshire might be mentioned. Such beds, Mr Fleming remarks, are seen alternating with the regular strata which accompany the coal in several places of the Hilderstone Hills ; and greenstone covers beds of slate-clay and marl which rest upon common grey-limestone at Hillhouse and Wardlaw, on the northern extremity of the Bathgate Hills, and at Kirkton on the southern side ; and these beds of limestone are *conformable* with the beds in which the coal is included.. Beds of amygdaloid, with a basis of greenstone, likewise occur in the coal-field at Bathgate.

Basalt, properly so called, it may also be mentioned, does occur in West Lothian : it appears between the old Silver Mines and the little eminence called Binny Craig. It contains numerous imbedded grains of olivine, and rests on beds of trap-tuff containing crystals of augite. These beds, however, Mr Fleming is disposed to consider as belonging to the *newest flœtz-trap formation* of Professor Jameson, and as altogether unconnected with the coal-field.

WILLIAMS *the mineralogist.*—As the history of Mr Williams is very little known, a few particulars may be acceptable ; more especially as my acquaintance with his son *, enables me to state them with some degree of accuracy.

He was the son of a clergyman in Glamorganshire, South Wales, and was born about the year 1730. While a boy, he resided a good deal at the copper-mines of Anglesea, and thus very early acquired a taste for mining pursuits. Having reached manhood, he travelled into Scotland, and was fortunate enough to be employed by the Commissioners for Forfeited Estates in the survey of some of the extensive Highland domains placed in their hands. In this employment he spent a considerable number of years. It was in the course of this survey that his attention was particularly

---

* Mark Antony Williams, Esq; surgeon in Haddington,— who obligingly revised and corrected this notice of his Father.

attracted by the *vitrified forts* of the north of Scotland ; of which he afterwards published an account.

About the year 1770, he took a lease of a coalwork on the banks of the river Brora in Sutherland. The seam then worked, proved to be of very limited extent, and the coal was so sulphureous, that when taken on ship-board, to be conveyed to the towns situated on the Moray and Cromarty Friths, it was apt spontaneously to take fire. This speculation, therefore, proved unsuccessful * ; and was abandoned in 1774.

After this, Mr Williams was successively engaged in the working of the lead-mines at Wanlockhead ; and of the silver-mines at Silver Hills, West Linton. In these operations, he embarked the greater part of the capital he had previously acquired ; and unfortunately neither of the concerns proved successful. He afterwards for some time superintended the coal-works at Blackburn near West-Calder ; and in 1778, became overseer and factor at Gilmerton Colliery, belonging to Mr Baird of Newbyth. Here he remained above thirteen years. During this period, his fame as a miner very greatly increased ; so that he was consulted by most of

---

* The coal on the banks of the Brora, has, of late years, been again tried, by order of the Noble and enlightened proprietor, Earl Gower, and promises fair to be useful to the North of Scotland. In a new pit opened in 1811, at the depth of somewhat more than 200 feet, a bed of " hard caking coal," 3 feet 3 inches thick, was found, and almost immediately below it, a bed of " hard splent coal," 1 foot 4 inches thick. *Sutherland Report*, 1812.

the coal proprietors in Scotland, whenever difficulties presented themselves in the course of their mining operations. In this interval, he brought out his publications. In 1777, under the auspices of Lord Kames and Dr Black, he published a description of the vitrified forts in the north of Scotland, in a curious tract entitled, " Letters from the Highlands * ;" and twelve years afterwards, he produced his large work on the Mineral Kingdom † ; which, it is believed, met with but little attention from the public at the time, but which certainly laid the foundation for no small share of posthumous fame to its author.

About the year 1791, he left Gilmerton, and became engaged, along with the late Dr James Anderson, in conducting *The Bee*, a periodical work then published at Edinburgh, and which acquired considerable reputation.

He afterwards spent nearly two years in travelling through England with Count Zenobia, visiting all the great manufactures and mines of that country ; and he was at last induced to go to Italy with that nobleman. On his arrival, he set on foot the working of limestone,

R

---

* " Letters from the Highlands of Scotland, addressed to G. C. M. (George Clerk Maxwell) Esq. By John Williams, Mineral Surveyor," 4to.   Edin. 1777. Mr Creech was the publisher, and has still some copies in his possession.

† " The Natural History of the Mineral Kingdom. By John Williams, Mineral Surveyor, F. S. S. A.," in two vols. 8v Edin. 1789.

coal and ironstone on some of the Count's estates near
Verona: he was going on with these improvements,
when he was unfortunately seized with a typhoid fever,
which proved fatal in the end of the year 1797, when
he had entered his sixty-eighth year.

I cannot close this short notice without remarking, that the merit of WILLIAMS, as an accurate observer, and original thinker, is perhaps greater than
many mineralogists are aware. While he certainly had
not even heard of the Wernerian doctrines, he published,
in 1789, his opinion, that "water has been the agent
in the formation of the strata; and that all the phenomena which we behold upon and within the superficies
of our globe, have been produced by water." He holds,
that "it was water that brought and poured the ingredients of all the mineral ores into the cavities of the
veins, while those ingredients were in a fluid state;"
and that "dykes" are cracks or fissures filled with heterogeneous matters by water;—just the Wernerian doctrines of mineral veins, and of the veins called "dykes."
Further, he observes, that "coals, and their concomitant strata," occur in basin-shaped coal-fields, which
are unconnected with each other; thus making a remarkable approach towards distinguishing the *Independent Coal Formation* of the justly celebrated
Professor at Freyberg. The opinion, that beds of
whinstone, (meaning chiefly greenstone and basalt,)
are *unerupted* lavas, had gained ground in Scotland,
being supported especially by the late ingenious Dr
Hutton: Mr Williams, however, shews, that "basaltic strata spread as wide, and stretch as far in
the longitudinal line of bearing, as any concomitant

strata, being regularly placed among the others which form the solid superficies of our globe;—they have therefore been formed in the same way as the other strata :—the columnar and glebous figures which they sometimes exhibit, have happened from drying;—and it would be strangely absurd to imagine that burning lava could come in contact with coal without destroying it *." Besides, this essay on the Mineral Kingdom contains much valuable practical information, particularly on the interesting subject of coal-mining; while some unsupported speculations, concerning " the mutations of our globe," and similar topics, in which the author has chosen occasionally to indulge, can very easily be forgiven.

Basalt and coal in Mull.—In quoting Mr Jameson s account of the occurrence of coal, having a roof and pavement of basalt, in the island of Mull, Mr Daubuisson has fallen into a slight topographical mistake, which, however, it may be proper to point out, to save some future traveller from being misled. Mr Jameson does not say that this bed of coal is situated on the " north east side of Mull," as stated in Mr Daubuisson's Memoir, but only that it is situated on the north-east side of a promontory in that island. His words are : " We crossed Lochskriddan near its mouth, and walked along the shore, which is low and

---

* Mineral Kingdom, vol. ii. pp. 48., &c.; also Introduction, p. xliii. where some spirited remarks are introduced on the Huttonian Theory, which had been then newly announced.

basaltic, until we approach Artown, where it juts out into a promontory, which presents several very beautiful ranges of basaltic columns. Upon the north-east side *of the promontory*, we observed, immediately upon the shore, a stratum of coal, which has for its roof a mass of imperfectly shaped basaltic pillars; and its floor is also basalt. The stratum is about twelve inches thick *," &c. The fact is, that Artown, and the promontory here mentioned, instead of being on the northeast side of Mull, form the *south-west* extremity of the island,—the point nearest to Iona or Icolum-kill, so celebrated in the ancient history of the Hebrides and of Scotland.

Coal and basalt in Faroe.—The examples here alluded to by Mr Daubuisson were first noticed by Captain Born, in the Memoirs of the Natural History Society of Copenhagen †; and the Captain's descriptions are copied by Mr Landt, in his Account of the Feroe Islands ‡. The following successive beds, proceeding from below upwards, are enumerated as occurring in an islet which has seemingly been detached from the larger island of Myggenaes, the most westerly of the Feroes.

1. Small crystallized basaltes.
2. Black stone, full of cavities.
3. Red petrified clay.

---

* Mineralogical Travels, 4to, vol. i. p. 214.

† Skrivter af Natur Historie Selskabet. Copenhagen, 1793.

‡ Description of the Feroe Islands, by the Rev. G. Landt; translated from the Danish,—London, 1810. Pp. 17. & 75.

4. Bluish grey stone, with calcedony.
5. Compact dark blue stone, coarse-grained fracture.
6. Petrified clay.
7. Black stone, with cavities.
8. Petrified clay.
9. Fine-grained hard grey stone, like sandstone.
10. Basaltic stone.
11. Basaltic columns.
12. Grey hard rock.
13. Coal, thin bed.
14. Black clay.
15. Basaltes.
16. Porous stone.

It is easy to apply the Wernerian nomenclature to several of the rocks here alluded to; but it is rather difficult to guess at the synonimes of others.

Nos. 1, 11, and 15, must mean either columnar basalt or greenstone; probably the latter: for, Sir George Mackenzie and Mr Allan, who visited Faroe in the summer of 1812, found that the perpendicular columnar rocks of Faroe, as far as they had an opportunity of observing them, were greenstone; basalt occurring only in narrow veins, and there divided into small horizontal columns, similar to those at Plauischen-grund, near Dresden, already described in the Memoir, *suprà*, p. 70.

Nos 2 and 7, according to Sir George Mackenzie, may probably signify a vesicular porphyry, he having observed such a rock, of a black colour, to be common in the islands.

Nos. 3, 6, 8, 9, called " Petrified clay," " Red petrified clay," &c. may, he thinks, be interpreted trap-tuff. He had no opportunity of visiting Myggenaes; but interposed beds of trap-tuff, more or less decomposed, and of a reddish colour, were frequent in the other islands. The porous stone, No. 16, Sir George likewise considers as probably trap-tuff in a state of decomposition.

No. 4, we may conclude to be amygdaloid, as the celebrated Faroe calcedony occurs chiefly in amygdaloid having the same basis as the trap-tuff.

No. 13, *Coal.* This seems to be of the same nature with the coal in the Hebrides, described by Professor Jameson, in his Mineralogical Travels *.

The principal beds of coal in the Faroe Islands, however, occur in Suderoe, the most southerly of the group, as the name implies. Near Præstfield, Mr Landt informs us, " the lowest stratum of coal is from a foot and a half to two feet in thickness : above this is a thin stratum of blackish-grey clay interspersed with small stripes of coal : the uppermost stratum of coal, is laminated (bituminous wood ?), and seven inches in thickness ; over this, lies a second stratum of the blackish-grey clay, with small stripes of coal. The whole of these dip to the north-east, with the other rocks."

In Sir George Mackenzie's collection, I have seen pieces of the Faroe coal, which were sent to him from Suderoe, together with specimens of the rock which occurs immediately above and below the coal. The coal includes the brown-coal and pitch-coal of Werner ; and the accompanying rock seems to be greenstone, and Sir George received from the same place, specimens of bituminous wood.

---

* Vol. i. p. 214—221.; and vol. ii. p. 76—87.

The coal of Faroe, therefore, would appear to belong to the *second coal formation* of Werner, or that which is subordinate to the newest flœtz-trap.

I cannot help expressing my regret, that our very intelligent Scottish travellers, Sir George Mackenzie and Mr Allan, had no opportunity of examining either Myggenaes or Suderoe, which are exposed islands, not easily approached in blowing weather, but which are the only places in Faroe where coal is known to occur; more especially as their examination of several of the other islands seems to have led them to conclude, that the Faroe trap-rocks in general are of igneous origin *. It may be proper to add, that Sir George Mackenzie brought from Nalsoe and other parts of Faroe, and has deposited in the cabinet of the Royal Society of Edinburgh, some curious specimens, having the twisted or rope-like aspect described by Daubuisson as occurring in Auvergne, and which he has expressed by the name *scories cordées* †.   T.

NOTE S.—(p. 141.)

Alternation of Basalt with Limestone, Sandstone, and Wacke.

THE alternation of basalt with limestone, sandstone, and wacke, in the island of Eigg, one of the Hebrides, is described by Mr Jameson in the 2d vol. of his Mineralogical Travels.

---

* Transactions of the Royal Society of Edinburgh, vol. vii.
† *Journal de Physique*, vol. lviii.

" The rocks which skirt the eastern part of the island, are very high, forming cliffs, and so well exposed, that we can observe in the most distinct manner the super-position, or the order and nature of the beds which compose them, from the surface of the sea to their summit.

These beds are nearly horizontal, and proceed from below, upwards, in the following order :

1. *Shistose-clay:* a few inches thick, and containing well preserved shells.
2. *Compact limestone :* a very thin bed.
3. *Shistose-clay :* about ten inches thick.
4. *Compact blue limestone :* containing shells, and of the same thickness as the preceding bed.
5. *Shistose-clay.*
6. *Basalt.*
7. *Fibrous limestone :* thin, and impregnated with bituminous matter.
8. *Basalt.*
9. *Argillaceous sandstone :* this sandstone constitutes an enormous bed, in which are contained several beds of basalt, from four inches to several feet thick : many impressions of shells occur in the sandstone.
10. *Basalt in columns ;* forming a very thick bed.
11. *Wacke ;* constituting a vast bed, and frequently containing fine crystallizations of zeolite.
12. *Red-wacke ;* in a very thin bed.
13. *Basalt,* forming a great bed.

All these beds are traversed in several places, by veins of basalt."—*Mineralogical Travels,* &c. 4to. vol. ii, pp. 36—40.

I again put the question, Whether these beds of basalt, some of which are only a few inches in thickness,
and alternate with sandstone containing petrifactions,
can possibly be considered as the remains of streams of
lava ?

~~~~~~~~~

NOTE T.—(p. 142.)

Basalt alternating with Shelly Limestone.

It may be proper to give Dolomieu's own account of
the appearances here alluded to, as detailed in the
Journal de Physique, vol. xxxvii.

" On the sides of the granitic hills of Auvergne, the
lavas alternate with calcareous rocks containing im·
pressions of shells, and are often covered by such rocks.
In the Vicentin and the Tyrol, there are calcareous
mountains, composed of horizontal beds, to the height
of more than 400 fathoms ; and in some instances twenty beds of lava, may be observed alternating with as
many layers of calcareous matter.— In Sicily, likewise,
about twenty regular alternations of beds of volcanic
rock and of limestone may be seen, uniting to form considerable mountains."

I think myself entitled to regard as basalt, the substances which Dolomieu calls lavas and volcanic rocks,
because he uniformly gives such names to true basalt ; and indeed, from what we know concerning

the localities of these rocks, they must be considered as basalt. Further, it was while Dolomieu was making his earliest observations, that he saw in Italy the appearances here described. The Sicilian naturalists are in use to talk of all black mineral substances as volcanic productions ; and it seems nowise improbable, that, at that early period, the comparatively inexperienced Dolomieu might implicitly adopt their nomenclature.

~~~~~~~~~~

NOTE *u*.—(p. 150.)

Puys of Auvergne and Vivarais.

MR DAUBUISSON states in the introduction to his Memoir, that while he is satisfied that the basalts of Saxony have had an aqueous origin, he " does not pretend to decide the question as to basalt in general *." It must be confessed, however, that, towards the conclusion †, he discovers an inclination to generalize the the neptunian doctrines, although this is done only in the way of conjecture, and with great modesty. It was, therefore, recommended to him by the National Institute ‡, (several active members of which are

---

* Introduction, p. 8.

† *Anteà*, pp. 154. and 167.

‡ Report to the Institute, by MM. Haüy and Ramond, subjoined to Daubuisson's Memoir.

keen Volcanists,) to visit the range of conical hills in Auvergne and Vivarais, and the extensive basaltic rocks which accompany them. He had already expressed a wish to undertake such a task *, and accordingly accomplished this mineralogical tour in the year 1804 ; travelling, as he had done in Saxony, on foot, with his hammer in his hand, and without any guide but a map and a compass.

The result of his investigations seems to have been a conviction, that the basaltic rocks of Auvergne are really of igneous origin;—in short, that the *puys* of that country are extinct volcanoes, and that the rocks in question are the lavas which flowed from them. While, however, he establishes the volcanic character of the *basaltic lava* of Auvergne, he is very far from intimating any change of opinion concerning the aqueous origin of the *basalt* and *greenstone* of the north of Germany. This seems to be the state of the fact, as may be gathered from an abstract, published in the *Journal de Physique* †, of a paper which he read before the Institute, on returning from his journey.

This paper, we are told, was divided into five sections. In the 1st, he gave an account of the mineralogical structure of the province : in the 2d, he described the chain of *puys* or conical hills, about sixty in number : the 3d and 4th contained descriptions of Mont Dor, and Cantal, lofty mountains differing re-

---

* *Suprà*, p. 150.

† Vol. lviii, for 1804, p. 310 ; also p. 422.

markably in character from the smaller hills; and, in
the last section, the author drew a comparison between
the basalts of Saxony and those of Auvergne.

It unfortunately happens, that on this last section,
(the most interesting in this place,) the abstract in the
Journal de Physique is much less full and satisfactory,
than on any of the others. But Mr Daubuisson's sentiments may be collected from some incidental expressions : I shall therefore give a condensed account of
the whole paper, taking notice particularly of these
expressions.

The fundamental rock *(sol)* of Auvergne, is common
granite, composed of felspar, quartz, and mica. In the
western part, it is covered with gneiss and mica-slate,
containing metallic veins ; and, in some places, with
greenstone-slate. In the valley of the Allier, limestone
appears, containing siliceous concretions, and much
impregnated with bitumen. In the low district, where
the Alagnon falls into the Allier, a coal country occurs ; every where else, the granite serves as the immediate support of the volcanic hills.

The chain of puys extends for above twenty miles.
Most of them stand detached. Their form is generally that of a truncated cone ; their sides are inclined at
an angle of about 30° ; and on their summits there is
generally a cup-like depression or crater, which is sometimes 300 feet deep. Their general elevation is from
900 to 1300 feet above the plain : the central and highest, the Puy de Dôme, being near 2000 feet.

The substances which chiefly compose these hills, are
described under the titles of Spongy Scoriæ, which are

often of a bright red colour; Twisted or rope-like scoriæ, *(scories cordées)*; Lapillo, or volcanic gravel; Vesicular or blistered lavas *(laves boursoufflées)*; and Basaltic lava, *(lave basaltique.)* In most cases these substances are confusedly heaped together; but the Puy de Dôme, and two or three others, form exceptions.

The sides of most of the puys are covered with herbage and brushwood; a few are wooded. On the top, as already mentioned, is generally found a crater, which is sometimes broken down on the side from which the lava had flowed. One of the most perfect examples, is the Puy de Pariou; and Mr Daubuisson's description of this is given at full length in the *Journal.* Its crater is so entire, and so nicely formed, that it appears as if " turned on a loom;" and the stream of lava, 200 feet broad, is observed, after having flowed for some way, to divide into two, the smaller currents then making their way to the low grounds in different directions, to the distance of four or five miles. These lava streams receive the general denomination of *cheires* from the inhabitants of the country. They commonly form a most barren and ungrateful soil: the cheire that proceeds from the Puy de Come, is so remarkable for its sterility, that it is called *Cheire de l'aumone.*

The surface of the lava is generally blistered and extremely rough, some of the asperities projecting several feet. An idea of the roughness of its surface may be formed, by supposing it to resemble that of melted lead poured into water. The interior of the lava is of a greyish-black colour, of a fine compact grain, hard, brittle, and frangible, differing from common basalt in these last characters. It contains crystals of augite and of felspar, the latter having a vitreous aspect, and rarely grains of olivine. The lava of the Puy de la Vache, however, very much resembles common basalt. It is vesicular at the surface, but fine-grained or compact in the interior or at the bottom.

The cheire of Volvic, is in some respects an exception. Its substance is of a lighter colour, and so soft, that it is easily worked. Quarries have therefore been opened in it, and it affords a pretty good building material. It contains pieces of quartz altered by the heat; and the walls of the cavities which occur in it, are set with spangles of specular iron-ore. This cheire of Volvic had evidently flowed from the puy of Nugere. A spectator standing on the summit of this hill, has a distinct view of the stream of lava which had issued from the crater beside him. He perceives that, in descending to the low ground, it had spread over a large basin, bounded by rocks of granite, and had surrounded some detached rocks which interrupted its progress. The tops of these detached granite rocks, are seen rising above the surface of this lake of lava.

" When I thus perceived (says Mr Daubuisson) that these lava streams had avoided obstacles, and obeyed all the laws of hydrodynamics, I could not help reflecting, that if the infringement of these laws by the basalts of Saxony had formerly influenced my opinion against admitting *their* igneous formation, the exact obeying of the same laws in this case, ought to produce an opposite conviction in regard to *those of Auvergne*."

It appears therefore, as already mentioned, that if Mr
Daubuisson was formerly inclined, from analogy, to consider *all* basalts as of aqueous origin, he has in so far altered his views, that he now excepts *those of Auvergne*. This
seems to be the extent of any change of sentiment; and it
is without reason that some incautious writers have stated, that Mr Daubuisson has entirely altered his opinion
concerning the origin of basalt in general.   His opinion,
stated in the most favourable way for the Volcanists,
may be supposed to amount to this, That a substance
nearly identical in character and nature, may have been
produced either from fusion or from solution.   He declares that some of the " *laves basaltiques*," or " *laves de
nature basaltique*" of Auvergne, bear the most striking
resemblance to the " *basalte ordinaire*" of Germany ; but
it may be remarked, that even the difference of nomenclature indicates that he considered them as of different
origin.

Mr Daubuisson seems perfectly aware of the difficulty
of accounting for the production of lava in a granite
country, and where the volcanic hills rest immediately on the granite.   The granite itself could not afford
the lava ; for this contains from 15 to 20 *per cent.* of
iron ; while the granite contains scarcely any.   The
rocks which afforded the lava, must therefore necessarily be situated below the granite.   Yet the granite
is, in some of the valleys, observed to be more than
1200 feet thick.

Caloric must, in one way or other, have produced
the fusion of the rocks, and raised the lavas to the craters ; but how that caloric was generated, where no

store of coal or other inflammable matter could exist, is a very difficult problem.

The lavas of Auvergne, Mr Daubuisson remarks, are of a date more remote than human history or tradition; yet they are recent in comparison with the great revolutions which the surface of the earth has undergone. They must, for example, have been subsequent to the excavation of the valleys; for they flowed into these, and occupied the lowest parts of them.   T.

NOTE X.—(p. 168.)

Volcanic Theory of Dolomieu.

A concise but distinct account of his peculiar views was given by Dolomieu in one of the latest of his papers, read before the Institute in 1797 *.

He supposes that the globe of the Earth is hollow. Its interior contains a sort of matter which is constantly in the state of a viscid paste. On this the inhabited land rests. The softness of the matter, if it is owing to heat, does not depend on heat of the nature of that produced in our furnaces. If heat assists in the operation, it must be by means of a certain something which tends

---

* *Journal des Mines,* No. 41.

to separate the particles, and to enable them to slide on each other,—a *vehicle* of fluidity different from caloric. The volcanic agents are elastic fluids, which raise up this soft matter, and, piercing the crust of the globe, place it on the surface. The matter does not undergo igneous fusion excepting when it happens to come in contact with the atmospheric air. Portions of this viscid paste, thus raised up by the disengagement of elastic fluids, have in various places spread themselves on the surface of the earth, and form basaltes of different kinds —Such is the hypothesis of Dolomieu.

We may remark, that his opinion, far from being favourable to the doctrine of those who regard every basaltic mountain as an ancient volcano, strongly militates against it ; for we see that Dolomieu, who had so carefully investigated volcanoes, and basaltes, felt himself obliged to abandon that doctrine, and to adopt the hypothesis now stated.

Dolomieu's opinion is liable to various strong objections.

1. That the globe of the earth is hollow, and that its centre is in a fluid or semi-fluid state, are gratuitous suppositions, not even rendered probable by any observations of philosophers.

2. Caloric is the only means known in nature by which stony substances can be held in a state of fluidity analogous to fusion. Any different, though similar agent, or even any modification of caloric, is quite unknown to us, and cannot be admitted.

3. Supposing the disengagement of elastic fluids from the soft matter in the centre of the globe, to produce an expansive force sufficient to overcome the resistance

offered by the thick and solid crust, the effect must
have been a vast rending of the crust, the parts which
opposed least resistance being shattered and dispersed.
Certainly these fluids would not escape by mere augre
holes!—for I may be pardoned such a comparison,
when the trifling openings through which volcanic eruptions take place, are compared with the extent of the
Earth's surface.

4. To produce an eruption by a new opening,—at
once to burst the crust of the earth, and to propel the
viscid paste,—the expansive force must be mighty indeed. It may well be asked if the equilibrium would
be re-established by the emission of some thousand
cubic yards of lava; for an eruption of Vesuvius or
Etna produces no more.

All nature may seem on fire to the eyes of the man
who has fled for his life before a current of red-hot lava, and his imagination will for a long time present
him with nothing but burning mountains. But, to the
dispassionate observer, who coolly surveys the surface of
our globe, Etna, Vesuvius, and all the volcanoes,—nay
all the basaltes that are known, taken together,—are
mere trifles, bearing an infinitely small proportion to
the whole mass of our planet. The products are nowise proportioned to the vastness of the cause invented
to account for them.

Without insisting on many objections which might
be urged against the hypothesis of Dolomieu, I shall
only remark, that the important influence which he as-
signs to volcanic eruptions, and which one would think
must every moment threaten the very being of our
globe, seems only to have been exerted on a single oc-

casion,—long after the surface of the earth had acquired its general configuration, and after organic beings
had begun to exist; for basalt is almost always found
above the other rocks, and it there appears in detached
portions, like the broken remains of a vast bed.   It often rests on alluvial soil, on coal, and on limestone containing shells.   It is true that basaltic rocks are sometimes observed covered by other mineral substances;
but such instances are rare, and the covering beds are
either sandstone or limestone of the newest formation;
and their occupying this position in regard to basalt,
does not necessarily imply that the basalt must be of
great antiquity.

Dolomieu appears to have held, that the fire of volcanoes, or the level of the fluid matter which by its
eruptions forms lavas, must be at a great depth below
the surface of the earth.   The observation of the depth
of the furnace in actual volcanoes, never could have led
him to such a conclusion. Spallanzani, in the first volume
of his Travels in the Two Sicilies, mentions his having
stood on the brink of the crater of Etna, and been able
distinctly to discern the matter in fusion at the bottom of
the abyss, at the depth of about 270 yards, or the sixth
part of a mile : " I can affirm, (he says,) that I distinctly saw a red-hot liquid matter, having a slight un-
dulatory and boiling motion.   Stones thrown upon it,
produced that kind of noise which indicated a viscid
paste." The volcano, it will be observed, was then in
a state of rest; and we here find the level of the fluid
matter contained in the volcanic caverns, to be 900 feet
below the summit of a mountain, which rises 9000 feet
above the level of the sea, and not less than 6000 above

the low grounds at its base. It is evident, therefore, that the volcanic furnace here exists in the body of Etna itself, and not at a great depth in the interior of the earth.

Dolomieu, however, may after all have been led to his conclusion by a strictly logical deduction, though it here proves erroneous. Basalt in all countries has very much the same character and appearance; it is almost the only substance which has, in different countries, been considered as a lava. If the basalts of different countries, therefore, be really volcanic productions, the fires of these volcanoes must all have been situated in the same kind of rock or material. But we know of no such general rock or material; and the numerous fragments of limestone, and the garnets, hornblende, augite, &c. found among the ejected masses, sufficiently shew that the volcanic fires are really situated in rocks perfectly analogous to those with which we are familiar.

FINIS.

Printed in the United States
By Bookmasters